Mountain, Water, Rock, God

Mountain, Water, Rock, God

Understanding Kedarnath in the Twenty-First Century

Luke Whitmore

UNIVERSITY OF CALIFORNIA PRESS

University of California Press, one of the most distinguished university presses in the United States, enriches lives around the world by advancing scholarship in the humanities, social sciences, and natural sciences. Its activities are supported by the UC Press Foundation and by philanthropic contributions from individuals and institutions. For more information, visit www.ucpress.edu.

University of California Press
Oakland, California

Suggested citation: Whitmore, L. *Mountain, Water, Rock, God: Understanding Kedarnath in the Twenty-First Century*. Oakland: University of California Press, 2018. DOI: https://doi.org/10.1525/luminos.61

Library of Congress Cataloging-in-Publication Data

Names: Whitmore, Luke, 1973- author.
Title: Mountain, water, rock, god : understanding Kedarnath in the
    twenty-first century / Luke Whitmore.
Description: Oakland, California : University of California Press, [2018] |
    Includes bibliographical references and index. | Whitmore, Luke 2018
    This work is licensed under a Creative Commons CC-BY-NC-ND license.
    To view a copy of the license, visit http://creativecommons.org/licenses |
Identifiers: LCCN 2018024588 (print) | LCCN 2018027937 (ebook) | ISBN
    9780520970151 (e-edition) | ISBN 9780520298026 (pbk.)
Subjects: LCSH: Ecology—Religious aspects—Hinduism. | Siva (Hindu deity) |
    Kidarnath (Temple : Kedaranatha, India) | Natural
    disasters—Religious aspects—Hinduism.
Classification: LCC BL1215.N34 (ebook) | LCC BL1215.N34 W45 2018 (print) |
    DDC 294.5/35095451—dc23
LC record available at https://lccn.loc.gov/2018024588

26   25   24   23   22   21   20   19   18
10   9   8   7   6   5   4   3   2   1

CONTENTS

ILLUSTRATIONS

FIGURES

MAPS

ACKNOWLEDGMENTS

This work came into the world with the help, support, guidance, knowledge, and advice of so many other people that in certain ways it feels strange to count myself as the author. I am deeply grateful for the guidance of many great teachers. In particular I would like to thank my PhD adviser Laurie Patton for her unflagging support and inspiration. She possesses the brilliant and special ability to respond to and elevate the work of others by showing them the original ideas already latent in their own work. I would also like to thank the other members of my committee: Paul Courtright, Joyce Flueckiger, and Don Seeman. I have learned and continue to learn a great deal from each of them; collectively I was truly fortunate to work with a dissertation committee who were so deeply insightful and intellectually generous in the guidance they gave and continue to give. I am indebted to the many fabulous teachers at the Landour Language School, the American Institute of Indian Studies-Jaipur Hindi program, and the American Institute of Indian Studies-Pune Sanskrit program for their fantastic language instruction.

It must be acknowledged that without the cooperation, hospitality, and guidance of many people in Garhwal this work would be nonexistent. Some of those people are no longer living. Here is a necessarily partial list: Sandeep Bartwal, Vijay Ballabh Bhatt, Mritunjay Hiremath, Brahamacari Jai, Bablu Jungli, Harshvardhan Kapruwan, Lokesh Karnataki, Madhav Karnataki, Rajesh Kumar Kotiyal, Darshan Lal, Pashupatinath Lalmohariya, Pradeep Lalmohariya, Gangadhar Ling, Shankar Ling, Shri Jagadguru Bheemashankar Ling, Bhupendra Maithani, Shivprasad Naithani, Deepak Negi, Vijay Negi, Anand Posti, Srinivas Posti, Ram Prakash Purohit, Bharat Pushpwan, Bhupendra Singh Pushpwan, Biru Singh Rana, Jitendra Singh Rana, Baghmbar "Shailu" Singh Rawat, B. R. Semwal, Om Prakash Semwal,

Sridhar Prasad Semwal, Umadutt Semwal, Pankaj Shastri, Shiv "Shibu" Singh, J. P. Shukla, Omkarnath Shukla, Bhagavati Prasad Tiwari, Tej Prakash Trivedi, Dhruv Vajpayee, Neelkanth Vajpayee, and Shankar Vishwanath, as well as many members of the Badrinath-Kedarnath Temples Committee and the Kedarnath Tirth Purohit Association and numerous others throughout the Kedarnath valley system, particularly in Ukhimath, Guptkashi, Gaundar, and Lamgaundi. A complete list would run into the hundreds if not thousands.

I would also like to thank specifically the following individuals who have both personally and professionally contributed in substantial and invaluable ways to the completion of this work as friends, teachers, and intellectual interlocutors: Amy Allocco, Connie Anderson, Michael Baltutis, Gil Ben-Herut, Nadine Beraradi, Rakesh Bhatt, Rikhil Bhavnani, Noah Bickart, Jessica Birkenholtz, Edwin Bryant, Ritodhi Chakraborti, Christopher Key Chapple, Corin Colding, Lisa Crothers, Maggie Cummings, Ian Curran, Corinne Dempsey, Antoinette DeNapoli, Spencer Dew, Diana Dimitrova, Tara Doyle, Lalita Du Perron, Diana Eck, Jennifer Saunders Forman, Susan Friedman, Parashar Gaur, Asher Ghertner, Radhika Govindrajan, Jonathan Greene, Udi Halperin, James Hare, Ann Harris, Josie Hendrickson, Arthur Herman, George Alfred James, Stephanie Jamison, Aftab Jassal, Tori Jennings, Paul Dafydd Jones, Maheshwar Prasad Joshi, Samantha Kaplan, Daniel Kapust, Advaitavadini Kaul, Molly Kaushal, Alice Keefe, David Koffman, Nirmila Kulkarni, Jon Levenson, Shanny Luft, Jim Lochtefeld, Preetha Mani, Brad Mapes-Martins, Sara McClintock, Afsar Mohamed, Rahul Bjorn Parsons, Kimberley Patton, Rafi Peled, Brian Pennington, Andrea Pinkney, Karin Polit, Neil Prendergast, D. R. Purohit, Madelyn P. Ramachandran, Rakesh Ranjan, Mani Rao, Keith Rice, Deborah Roberts, S. Brent Rodriguez-Plate, Leah Anderson Roesch, Jordan Rosenblum, Joseph Russo, William Sax, Pankaj Semwal, Megan Sijapati, Rana P. B. Singh, Fred Smith, Abigail Sone, Shashank Srivastava, Padmanabhan Sudevan, Urmila Thapliyal, Roy Tzohar, Peter Valdina, Phillip Webb, Michael Witzel.

I would like to thank those whose generous funding and logistical support contributed to the work of this dissertation. The Fulbright Student Program and the Sinclair Kennedy Traveling Fellowship funded my first research trip to India in 1999–2000. The American Institute of Indian Studies provided me language-study grants and predissertation research funding in 2004–5, funded my dissertation fieldwork from 2006 to 2007, and graciously provided logistical support on other occasions. I am particularly grateful to Elise Auerbach of AIIS-Chicago and Purnima Mehta of AIIS-Delhi. The Laney Graduate School of Emory University provided funds at several key moments, including a SIRE fellowship to complete my dissertation. I also thank the Indira Gandhi National Centre for the Arts for hosting me during my dissertation fieldwork and then later my work as a research scholar, and HNB Garhwal University (Srinagar) for providing an extraordinarily collegial research environment at many different times over the years.

I gratefully acknowledge the support of both the Office of Research and Sponsored Programs at the University of Wisconsin-Stevens Point and the Institute for Research in the Humanities at UW–Madison for their support through a UW System fellowship for a year at the IRH. I thank the South Asia Center at the University of Wisconsin–Madison for their collegiality and support. I would also like to give special thanks to Susan Friedman, who directed the IRH at UW–Madison during my fellowship year there in 2014–15. Her personal guidance and her presence at the Institute were critical for finding the voice I achieved in this book. I also gratefully acknowledge support for the publishing of this work that came from the Office of Research and Sponsored Programs at UW–Stevens Point. I owe a great debt to my colleagues in the Philosophy Department and on the fourth floor of the Collins Classroom Center more generally at UW–Stevens Point for their logistical, psychological, and intellectual support at key moments during the researching and writing of this book. I gratefully acknowledge the help of Keith Rice and the UW–Stevens Point Geographic Information Systems Center in preparing the maps for this book. I would also like to thank the editorial team at University of California Press, particularly Eric Schmidt, Maeve Cornell-Taylor, Kate Warne, and Elisabeth Magnus, for their fantastically competent work on this book and for their enthusiasm for and thoughtful engagement with this project. I am grateful to Cynthia Col for her thoughtful preparation of the index.

I would like to thank my wife Judith and my daughter Talya for their support and patience with me during the long process of researching and writing this book. I could not do this without the two of you. *Talya Ima Abba.* I would also like to thank my parents David and Lois Whitmore and my sister Flora for supporting me and showing me the way as I learned to trust my own creativity. Thank you to the Sones collectively for their constant support and to Abigail Sone in particular for her fantastic editing and Jonathan Greene for his unstinting intellectual support.

Notwithstanding my gratitude to all those who have helped me in the production of this work, any faults or errors found in this work are wholly my own responsibility.

NOTE ON TRANSLITERATION

This book contains words, phrases, and passages that are translated from Hindi, Sanskrit, and occasionally Garhwali. In English-language scholarship on South Asia there is no universally agreed-upon system of transliteration and translation for this particular multilingual translation situation. I have therefore made a set of choices that both keep the work accessible to readers without a background in South Asian languages and at the same time provide linguistically inclined readers with a bit more information where it is especially relevant.

In the text itself I eschew diacritics and instead render the word in plain roman transliteration in a way that most closely approximates its pronunciation and indicates the specific language being used: for example, "They are writing" (Hindi: Ve likh rahe hain). If a specific word or proper name already has a known and common convention of use in English, such as Shiva or *linga* or *Ramayana*, then I use that spelling. On occasion I will add an English plural suffix to a Hindi or Sanskrit word (e.g., Hindi: *yatri*, meaning "pilgrim/traveler" would become *yatris* after the English plural suffix "s" has been added). For transliterated words that exist in more than one language I adjust the default language of reference between Sanskrit and Hindi (e.g., *Mahabharata* as opposed to Hindi: *Mahabharat*) according to my own judgment of what best serves the multiple audiences of this work. Bibliographic information will, where appropriate, be given with diacritics. Where they are particularly relevant I provide transliterations of longer passages in Hindi and in Sanskrit in the endnotes in which diacritics are employed. Unless otherwise mentioned, all translations are my own.

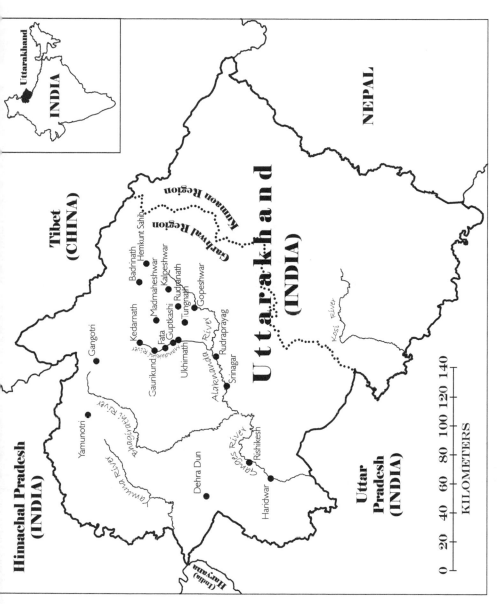

MAP 1: Uttarakhand. Courtesy of University of Wisconsin-Stevens Point Geographic Systems Information Center.

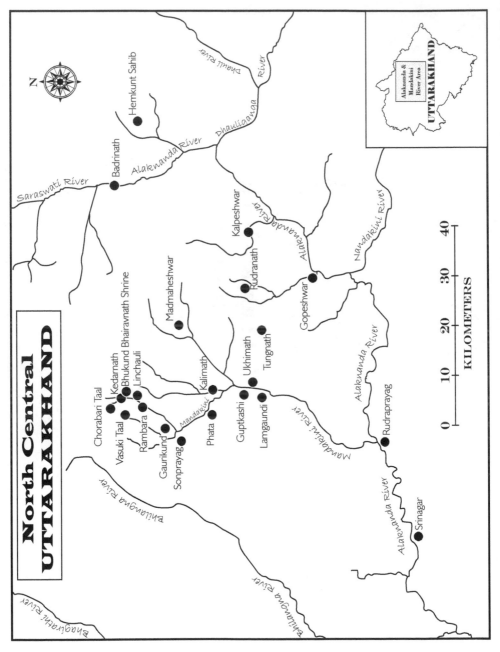

MAP 2. Uttarakhand North Central Region. Courtesy of University of Wisconsin–Stevens Point Geographic Information Systems Center.

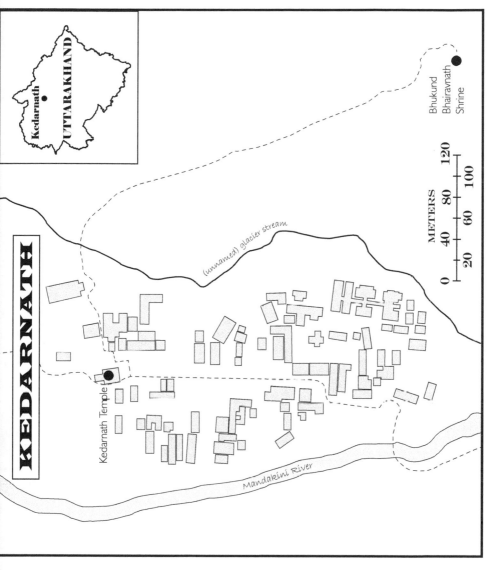

MAP 3: City of Kedarnath. Courtesy of University of Wisconsin-Stevens Point Geographic Information Systems Center.

Introduction

In the Direction of Kedar

When one had almost reached the Hindu shrine of Kedarnath, there used to be a point along the footpath following the Mandakini River where, in good weather, the top of the temple came into view against the backdrop of a bright, wide Himalayan panorama. In the related languages of Hindi and Garhwali, such vantage points bear the title of "Sight-of-God" (Hindi: *devdarshini, deodekhni*). *Devdarshini* is the point from which you are first able to glimpse the deity of a particular place, in this case Shiva in his Kedarnath form. Before June of 2013, *devdarshini* was located above the western bank of the Mandakini River, about two kilometers from the temple and twelve kilometers from the beginning of the footpath in the village of Gaurikund. The temple of Kedarnath lies on the valley floor at almost 3,600 meters above sea level, enfolded on three sides by mountains. Almost all travelers to Kedarnath, from perennial residents to first-time visitors to non-Indian regular visitors such as myself, used to stop at *devdarshini* to catch their breath, encounter the view, and take stock of the place where they were about to arrive.

Devdarshini often evokes a strong response. I once spoke with a woman from the Indian state of Maharashtra who had come to Kedarnath seven times.[1] She said that when she arrived at *devdarshini* she experienced what in Hindi is called *sakshat darshan*—a visual encounter with (or more technically a visual knowing of) the divine in its most true form. For her, *sakshat darshan* meant that from this point she saw the entire Himalayan panorama as the face of Shiva. On her pilgrimage to the region in the 1920s Sister Nivedita (1928, 39–40), a famous Anglo-Scot-Irish disciple of Swami Vivekananda, described the moment of *devdarshini* in this way: "At last comes the moment when the temple is visible for the first time. A shout goes up from our carriers and many prostrate themselves. We press forward,

FIGURE 1. Approaching Kedarnath.

more rapidly than before." The sight of *devdarshini* functions as a moment of relief and inspiration for visitors to Kedarnath, many of whom are finding themselves in the Himalaya and at high altitude for the first time in their lives as part of their performance of what is known as the Uttarakhand Char Dham Yatra (Hindi: Uttarakhand Four Abode Pilgrimage). *Yatra* is originally a Sanskrit word that could be reasonably translated as "pilgrimage." Today in many South Asian languages the term *yatra* possesses a semantic range that includes everything from "pilgrimage" to "vacation" to "tour." The Uttarakhand Char Dham Yatra is a journey to the abodes of four deities: Yamunotri (of the goddess/river Yamuna), Gangotri (of the goddess/river Ganga), Kedarnath (of the god Shiva), and Badrinath (of the god Vishnu). This journey involves travel by bus, car, foot, and pony over hundreds of kilometers of Himalayan terrain and is typically carried out in the short period of approximately twelve overwhelming, draining, transformative days. In recent decades performance of the *yatra* by helicopter has also been increasing in popularity—that is, in good weather. The Uttarakhand Char Dham Yatra is one of the most popular *yatra* journeys in North India today.

The sight at *devdarshini* of the white glaciers flowing down into the darkness of the lower mountains, divided by waters of the Mandakini River, which is one of the Himalayan tributaries of the Ganges, localizes the famous story of the descent

FIGURE 2. Approaching Kedarnath.

of the goddess Ganga in her river form into the world and underscores Shiva's relationship to the Goddess in her aquatic manifestations. Ganga descends at the request of King Bhagiratha, who has performed thousands of years of ascetic practice so that the souls of his ancestors may be purified through the presence of Ganga on earth. Ganga finally assents but says that her unfiltered power would be too much for the earth to bear. Thus a mere trickle of the full power of the Ganga exits the protective filter of Shiva's matted hair and enters our world. This panorama also evokes perhaps the most common act of ritual devotion to Shiva, the pouring of water over a *linga*. *Linga* is a difficult word to translate. Diana Eck (1999, 4) expresses it in this way: "the simple stone shaft that is the symbol of Shiva."[2] It is the complicated, fluid, indexical sign of Shiva's presence in and relationship to the world.[3] Shiva's presence in Kedarnath specifically is "self-manifest" (Sanskrit: *svayambhu*). He dwells, not in a form created by human hands, but rather in a form that was always already there. Further, his relationship to the devotee is not stable. In many ways, serious devotion to and worship of Shiva place the practitioner on the path to becoming a Shiva, a multiform of God. As Richard Davis (1991, 52) translates an important phrase of Shaiva (Shiva-oriented) ritual: "Only a Śiva can worship Śiva." The panoramic view, then, indexes not only the nature of the cosmos and the self-manifest nature of the god but also the personhood of the visitor who is taking in the view. It registers, both inwardly and outwardly, the fluid "oscillation" or "pulsation" that is the "ubiquitous principle of a dynamic universe, governing all creation" (42). The panorama of Kedarnath on a clear day fuses iconic, aniconic, and natural modalities for experiencing, worshipping, and becoming the conjoined presence of Shiva and Ganga in this world. Kedarnath is an especially storied place in the already legendary Himalayan region of Garhwal, famously known and marketed today as the Land of the Gods (Hindi: Dev Bhumi). Let us proceed to Kedarnath itself, the next step in exploring what the recent history of this complex, powerful place has to teach us about how religious worldviews frame human-nature relationships in the twenty-first century.

MEETING KEDARNATH

The village of Kedarnath, whose defining feature is the temple, lies on the valley floor at the upper edge of the Lesser Himalayan mountain range, quite a bit higher than the altitudes at which most Garhwalis live. Kedarnath occupies a distinctive place in the networks of modern Hindu pilgrimage destinations. It is difficult to reach. In recent decades the road ended in Gaurikund, about 1,800 meters in altitude and fourteen kilometers in distance short of the temple. Yet at the same time Kedarnath had become a major destination. Approximately half a million people visited Kedarnath in 2007. It is not in the same category of inaccessibility as

FIGURE 3. The Kedarnath temple.

FIGURE 4. Panorama of Kedarnath village from above the Bhukund Bhairavnath shrine.

a journey to the famous Mount Kailasa, or even the arduous journey to Amarnath, where Shiva dwells in the form of an ice *linga*. But it has remained a Himalayan pilgrimage destination evocative of the difficulty of premodern *yatra* to the Himalaya even as the number of its visitors has risen.

Kedarnath is a "crossing-over place" (Sanskrit: *tirtha*)—a place that offers the possibility that one can "cross over" or "ford" the ocean of rebirth. It is a place that provides special access to that which is underneath and over and within and beyond. It is a place that grants wishes, heals, and purifies *karma*. It is a *dham,* an abode or dwelling place of a deity. Kedarnath is best viewed primarily as a *dham,* an especially powerful abode of Shiva. More specifically, along with Manimahesh in modern-day Himachal Pradesh, Amarnath in Jammu and Kashmir, Kailash Mansarovar in Tibet, and Pashupatinath in Nepal, Kedarnath is one of the most famous places where Shiva is understood and experienced to dwell in the Himalaya, the mountain range that is famously both his father-in-law and his preferred residence, the place where he can pursue his yogic practices in natural solitude. It is one of twelve *jyotirlingas,* twelve locations spread throughout India where Shiva's universal form of a vibrating column of light embeds itself in the ground.[4] The *linga* contained in the temple is the focal point of Shiva's presence in Kedarnath. It is *svayambhu*—an a priori presence in the form of a roughly triangular piece of mountain rock that emerges from the ground and that muddies the relationships among god, place, and cosmos. This dwelling happens inside and outside the temple.

Kedara is a Sanskrit word that means marshy ground, soil mingled with water. It is usually glossed in Hindi with the term *daldali bhumi,* or swampy land. It occurs when and where snowmelt, rain, and river water turn the ground into a marshy ooze of varying consistency. Thus the name Kedarnath (Kedaranatha or Kedareshvara in Sanskrit) literally means "Lord of the Marshy Ground." It is both a place-name and a name for Shiva. The entire Kedarnath end-valley, starting from near *devdarshini,* lies buried in snow for approximately half the year. The snow melts and the end-valley is snow-free for most of the summer pilgrimage season, with the exception of the occasional snowstorm. For travelers on the path the most common greeting is "Jai Kedar" or "Jai Shri Kedar," a greeting and naming that addresses both the place and the god. It is one of many signals that Shiva is felt both inside and outside the temple—a theme that runs through most if not all aspects of the place of Kedarnath.

Old stories about Kedarnath enshrine this location as a point of transition and exit out of the human world. It is one of the places associated with the Climb to Heaven (Hindi: *swargarohan*) of the five Pandava princes and their joint wife Draupadi. Desperate to cleanse themselves of the *karma* generated during the massive civil war narrated in the *Mahabaharata* epic, the Pandavas head to the Himalaya as renunciants, leaving the political world behind. They walk up into the Himalaya and thence to heaven on what is known as the Mahapath/Mahapanth, the Great Path. The oldest brother, Yudhishthira, does so without leaving behind his physical body. This feat aligns itself with one of the central themes of tantric practice—liberation through the body as opposed to liberation as exit from the material world. The *Kedarakalpa* (Sanskrit: literally "account of Kedara"), a late tantric text of uncertain date, tells the story of five spiritual adepts who walk a version of the Great Path into the mountains that passes through the Kedarnath locale and beyond into the abode of Shiva (Goswamy 2013; Viśālmaṇi Śarmā Upādhyāy 1952; Padumā and Hajārībāg 1907). In the narrative worlds that touch on Kedarnath one finds a sense of the place as a limit point, a door to zones beyond.

THE FLOODS OF 2013

Recently Ganga gave the world a small taste of her fuller power that is normally held in check by Shiva as she descends into our world. The circumstances underscored the idea so clearly attested in the story of the Pandavas that Kedarnath functions as a place of crossing between life and death. In 2013, the monsoon came to Uttarakhand early and with special force. Intense rains resulted in widespread floods and landslides. Because it was the height of the summer pilgrimage season, over seventy thousand *yatris* were stranded, and by extremely conservative estimates hundreds of locals and *yatris* perished (Tripathi 2013; BBC News India Staff 2013a; Zee Media Bureau and Press Trust of India 2013). The floods and landslides prompted massive rescue and relief operations, a drama whose plot occupied (and continues to sporadically receive) national attention. Kedarnath was hit with two flood events, the second a truly dreadful wave of water and debris that resulted from the bursting of a natural debris dam on a snowmelt-fed lake located just above the village of Kedarnath to the northwest. By even the most conservative of estimates, hundreds of people in and around Kedarnath did not survive (table 1).

One early assessment of the impact of these flood and landslide events was carried out by Ravi Chopra (2014), director of the People's Science Institute in Dehra Dun. As this preliminary assessment makes clear, the events traumatized the region. Yet much of the trauma could have been avoided, for the impact of the flooding was multiplied by the lack of a well-developed disaster relief plan; insufficient sensitivity to meteorological warnings; insufficient acknowledgment of and planning for the clear, obvious likelihood of this kind of flood event; insufficient government regulation of the number of visitors; lack of alternate escape routes; unregulated construction of private and commercial building in floodplains and near riverbanks; lack of sufficient understanding about how human-built dams affect what happens to a river during floods that bring increased water volume and, more significantly, debris flow and muck; and the short-sighted widening of roads built in landslide-prone areas.

TABLE 1. Preliminary Statistics on the Uttarakhand Floods of 2013

Affected persons: 5 *lakhs* [500,000] (approx.)
Affected villages: 4,200
Severely affected villages: over 300
Persons injured: 4,463
Number of dead persons: over 900*
Number of missing persons: 5,748
Number of *pukka* [permanent stone/brick/timber-walled] houses damaged: 2,679
Number of *kuccha* [mud-walled] houses damaged: 681
Number of animals lost: 8,716
Number of roads destroyed: 2,302
Number of bridges washed away: 145
Number of drinking water schemes damaged: 1,418
Number of villages without power: 3,758
Preliminary number of landslides: 2,395

SOURCE: Chopra (2014).
*Chopra notes, "These are government figures. Most unofficial estimates of dead and missing are much higher."

In short, much of the impact of the floods and landslides would have been vastly mitigated by better planning, road and building construction, and regulation. Fewer would have died and been injured. The water and debris would have destroyed fewer buildings and roads. Travel to the region in immediately subsequent years would not have declined as far as it did, and locals would probably have invested more soberly and sustainably in commercial infrastructure before the floods and would have experienced less severe financial loss afterwards. As journalist Jay Mazoomdaar (2013) put it in his article "How Uttarakhand Dug Its Grave," "A state created to safeguard the hill people, has become a graveyard of pilgrims and local aspirations in just over a decade."

After June of 2013 Uttarakhand entered a highly public recovery mode, and Kedarnath has been one of the visible, important, and contested symbols of how Uttarakhand should move forward during this process. For visitors, the vast majority of whom were Indian, the scope of the disaster was a betrayal of expectations—the modern twenty-first-century nation-state should be able to give advance warning of events like this and be prepared to hit the ground running in the aftermath, particularly after the Indian Ocean Tsunami of 2004. *Yatra* in the Himalaya was supposed to have become a journey for the whole family. For Uttarakhandis the sense of betrayal and anger went deeper. The region that in 2000 became the separate state of Uttaranchal (and was later renamed Uttarakhand) has been marked since the late 1800s by contestation between local mountain residents and outside forces (the British colonial government, British timber companies, the Indian Forestry Department, the Uttar Pradesh state government from which Uttarakhand broke away in 2000) over the control of natural resources such as forests and rivers. It is the region that in the 1970s popularized the tree-hugging movement (known in Garhwal as Chipko) and saw the construction (also begun in the late 1970s) of what would be one of Asia's largest civil engineering projects, the Tehri Dam. Traditions of protest and resentment surrounding the exploitative extraction and control of natural resources during the colonial and postcolonial period ultimately created the conditions for the creation of a separate state that, according to Pampa Mukherjee (2012, 201), was founded on a "strong alternative vision of development" attuned to the human and ecological needs of the region.

This did not happen. Instead, much of the state, both inside and outside the state government, ignored what tourism studies would call the "carrying capacity" (Coccossis and Mexa 2004) of the region so that it could service the growing number of visitors to the region and the growing demand for electricity. For some, this disaster registered at least partially as the anger of Shiva at the de-sacralization of the region as it embraced ever-growing levels of pilgrimage tourism. For others, the tragedy was largely a human tragedy of poor planning, greed, and systemic miscalculation. For many, it was both. The *devdarshini* I knew has been washed away and a new path has been built on the eastern bank of the Mandakini. The village is now partially enclosed to the north by a massive set of protective walls. The temple courtyard now includes a massive boulder whose bulk protected the temple during the floods.

This is the reason that much of what I have written, so far, has been in the past tense. Much is different today, and everything that I saw before 2013 now, in hindsight, feels like foreshadowing. Even in good weather in Kedarnath there was always a sense of unpredictability, of an uncertainty built into the ground. Journey to this place of divine hyperpresence has never been without risk. The precarity of human presence in the challenging environment of the Himalaya, where earthquakes and landslides go with the territory, is something that has been fundamental for humans living in and visiting the Himalaya for millennia.[5] This reality reasserted itself with new force in the twenty-first century. Both locals and *yatris* often reminded me how difficult it was for anyone to reach Kedarnath. Finding oneself in Kedarnath for any reason was at some level a karmic triumph. During even a normal pilgrimage season the monsoon rains cause landslides that block the road, and visitors may find themselves stranded in a mountain valley whose modern infrastructure is suddenly revealed to be skin-deep, a tracery on the surface of the land. On June 21, 2013, as reports continued to come in about the number of dead in and around Kedarnath, I remembered a conversation I had once had with a Kedarnath pilgrimage priest (Hindi: *tirth purohit*) in which he remarked (my paraphrase) that Kedarnath was not a place where humans were meant to live.[6]

PLACE/DEITY IN CONTEXT

We can best understand the floods of 2013, particularly in the Kedarnath valley, by situating these terrible events within their broader ecological and religious contexts. The power and importance of Kedarnath are a tapestry woven out of many threads: place-based concerns about life in the Himalaya, transregional devotion to a famous god, tourism from across India and around the world, and the particular environmental history of the central Indian Himalaya. This environmental history includes two issues critical for life in the twenty-first century: deforestation and the control of water. The floods of 2013 and their aftermath rendered the interdependence of these threads terribly, publicly clear. We must try and learn what we can from this moment of awful transparency. The case of Kedarnath offers an opportunity for thinking through not only the nature of these interdependencies (which for many is already not surprising) but also the longer stories of how foundational factors structure this overall web of connection. And here we arrive at the central argument of the book—an argument for a way to image this interdependence and its implications. For centuries, the *enmeshing* of Shiva with the Himalayan environment has animated how Hindus conceptualize and experience Kedarnath. Shiva is both inside the temple and part of the Himalayan place itself. The fluid overlap of place and god frames, in important ways, the place as a whole. Over time this long-term pattern of the place combined with region-specific changes in land use, a postcolonial pattern of regulatory myopia about Himalayan environmental realities, and in recent decades a massive, rapid rise in the number of visitors to the region. It was at this point that the situation became a ticking bomb that encouraged short-term, selfish and self-motivated action by those with livelihoods connected to pilgrimage tourism in Uttarakhand. In important ways the terrible power rampant during the floods was not a departure from noncatastrophic times. Landslides, floods, and earthquakes are to be expected in the Himalaya—such power is always latent in the mountain landscape. What the aftermath of the floods laid bare was that in recent decades the situation in Uttarakhand made this well-known fact easy to forget.

It will be impossible for many years to visit or even discuss Kedarnath without beginning from the time of the disaster. However, what happened in 2013 is best understood in the broader context of how people have understood and experienced Kedarnath both in the recent past and in earlier times. In writing about the conditions leading up to and the reactions after a disaster in a place of religious significance, a balance must be struck between a commitment to an urgent critique that is rooted in the recent past and the necessity for placing that recent past in broader context. To address this analytic challenge I work with the idea of a place's "persistent patterns"—persistent not because the place has been frozen in time but rather because the pattern is the continually emergent result of particular and changing constellations of culture, environment, and political economy (F. Cho and Squier 2013, 360). This feature of the place, its "persistent patterns," functions as a point of commonality among people who find themselves there for a variety of different reasons: as tourists, as pilgrims, as laborers, as service providers, as guides, as shopkeepers, as priests, and as employees of the government. I argue for the importance of noticing a particular persistent pattern: a long-standing ambiguity about the specifics of how Shiva inhabits Kedarnath that has demonstrably informed experiences, conceptualizations, and representations at many different moments in the recorded history of this place. The terrible events of 2013 are part of the broader persistent pattern of Kedarnath and, when viewed in a certain way, help us see things about this broader pattern that might not otherwise be easily visible.

The disaster in 2013 publicly affirmed the ambiguous, fundamentally Himalayan and Shiva-oriented character of Kedarnath. At the same time, the floods made it clear that the patterns of commercialization, development, and regulation of recent decades in Uttarakhand, patterns that both had arisen in response to new statehood and an influx of middle-class pilgrims and tourists and were the continuations of policies first employed by the British in the nineteenth and early twentieth centuries, were starkly out of place. The floods underscored a persistent geological fact about life in the Himalaya that had been underemphasized in recent decades of road and hotel construction: floods, landslides, and earthquakes are an inescapable part of this mountainous, seismically active terrain. The sense of both consonance and dissonance with the Kedarnath (and more broadly the Garhwal Himalaya) of earlier decades and centuries is at once religious, economic, political, and ecological. People connected to Kedarnath today therefore understand both the disaster and the recent short-sighted development that multiplied the impact of the disaster both as the consequence of human disregard and as an indication of a growing disconnect with the Himalayan environment and its resident divine powers. As local and regional stakeholders took charge of the reconstruction of Kedarnath with a national audience in mind, religious understandings about Shiva, the Himalaya, nature, and the power of river goddesses infused the push for a more ecologically sound and economically viable brand of pilgrimage tourism and regional development. Uttarakhandis and visitors to the region for years to come will grapple with the short-term and long-term economic, cultural, and moral costs, both to themselves and to others, of the practice of pilgrimage tourism in the central Indian Himalaya. The floods of 2013 provoked a large-scale referendum on questions of appropriate development in the central Indian Himalaya. It is the task of this book to contextualize, in a Kedarnath-centric way, the beginnings of a new post-2013 chapter in the ongoing story of this region. Doing so will offer a portable, holistic model for thinking about how religious worldviews inform conceptions, experiences, and practices connecting to the natural world that can be deployed across multiple scales. Below I suggest that the idea of "the experience of place as an eco-social system characterized by complexity" is a useful way to conceptualize this situation as a whole. But this way of thinking about the Kedarnath situation does not fully describe or understand who and what is present in and around this place. What analytic framework could?

THE PURPOSES OF THIS BOOK

In 2010, when I had left Kedarnath and passed *devdarshini* going back down to Gaurikund, a commercial helicopter carrying *yatris* flew overhead. I was taking a picture of a *yatri* who had chosen to ride a pony up to Kedarnath rather than walk. The choice to ride a pony had in recent years become increasingly popular (Whitmore 2016). At that time, I had seen the presence of a helicopter overhead as the most visible evidence of how Kedarnath was becoming more and more of a destination for middle-class and upper-middle-class Indian *yatris*. Today, the sight of helicopters in the Kedarnath valley carries an additional association. Helicopters in 2013 were almost the only way that the government could reach those stranded by floodwaters and washed-out roads. For much of 2013 they were the only way to deliver supplies to the area, but often the weather conditions in the Kedarnath valley prevented even helicopters from making the journey. One of the most urgent priorities in the early reconstruction of postflood Kedarnath was the construction of landing pads capable of supporting the weight of heavy cargo helicopters.

As this place has changed, so too have my responsibilities as a scholar. This is intended to be a book about a *place* that shows how attending to the emplacement of religion helps us understand the relationships of ecology, disaster, and development. In many important ways, Kedarnath-as-place is the central character in this story. This approach emerges from numerous ethnographic encounters and my responsibilities to those about whom I write. During an eighteen-month field-work period in 2006–8 I spent as much time as I could in Kedarnath. This meant approximately six months during 2007 and three weeks of 2008. I have since visited Kedarnath briefly in 2011 and 2014 and the Kedarnath valley (but not Kedarnath) in 2017. The longest I spent in Kedarnath itself without respite, for reasons of health, logistics, and research, was about two months. During these periods of time I attempted to learn from and about as many different kinds of people as possible. I spoke with *yatris* from all over India and the world. I made sure to spend time with *tirth purohits* (Hindi: pilgrimage priests), employees of the Badrinath-Kedarnath Temples Committee, police, shopkeepers, and pilgrim rest house and lodge managers from different parts of the Kedarnath valley, contractors and cleaners from the plains, Nepali laborers, renunciants, and Western visitors. I spent most of the remaining months of that fieldwork period further south in the Kedarnath valley in Ukhimath, the administrative center for the division of the Rudraprayag district in which Kedarnath is located. This time was itself punctuated by trips to the British Library in London for information about Kedarnath in the India Office Library and to the Scriptorium of the Sanskrit Dictionary Project in Pune for references to Kedarnath in Sanskrit literature. While based in Ukhimath, I spent most of my time introducing myself to the three valleys of the Kedarnath valley system. I visited dozens of villages and important shrines in these three valleys so that I would understand how Kedarnath fit into its local contexts and so that I would have a connection to persons working in Kedarnath who came from all these places. I was able to carry out and record structured and semistructured interviews with Kedarnath and Kedarnath valley residents and several renunciants.

My overriding goals, formed by this time in Garhwal, have been to shape the book in a way that would allow a serious reader to imagine something of what it feels like to stand in Kedarnath at the beginning of the twenty-first century and at the same time see the broader and longer contexts that surround these recent decades. This has meant marshaling a number of approaches from across the humanities and social sciences in service of a very specific kind of descriptive holism. The goal of this description is to bring the reader partially into the community of those who have been to Kedarnath in a way that inspires long-term concern for the place and the region and the broader social, cultural, economic, political, environmental, religious, and ecological processes that flow through and center on Kedarnath. There is much to learn from the case of Kedarnath as we continue to orient ourselves to life in the twenty-first century, framed as it is by the increased flow of people and capital, shifting environmental realities, the persistent embedding of what some would call "religion" in how people understand themselves and others to be in the world, and the recognition that we are more a part of our world than many of us have realized.

This approach to writing about Kedarnath has been my response to the idea that, as Laurie Zoloth (2015, 374) puts it in her contribution to a special issue of the *Journal of the American Academy of Religion* on climate destabilization, "scholars in the humanities have duties to the world." Todd LeVasseur (2015, 297), in his introduction to the same special issue, poses this question: "What does it mean to be a religious studies and/or a biblical scholar professional, working as an employee at an institution of learning on a biophysical campus, on a planet undergoing rapid anthropogenic climate destabilization?" While the exact causal relationships between the floods of 2013 in Uttarakhand and the global phenomenon of anthropogenic climate change are not yet wholly clear (I discuss this more in the body of the work), it is very clear that much of the impact of the disastrous flooding was human caused, making Kedarnath a natural disaster with an "unnatural history" (Steinberg 2000), or, more bluntly, an *unnatural* disaster. But whatever the exact causal, scientific explanation may be for why the floods of 2013 happened in Kedarnath and across Uttarakhand in the way that they did, these events should be understood as exactly the kind of unnatural disasters that have already increasingly begun to mark the twenty-first century and that will inevitably disproportionately affect residents in the Global South. This is something that I have come to understand as part of my own journey in writing this book. I did not begin my career as a scholar of human ecology or even of what can now be called the subfield of ecology and religion. In recent years of thinking about this project I have followed my concern for what happened to people I know who live in the Kedarnath valley into these fields of study. I have arrived at the rather unsurprising conviction that the better we are able to understand the relationship between religious worldviews and the ecosystems of which humans are but a part, the more likely it is that particular forms of human action and environment-related policies of land use, consumption, investment, regulatory oversight, and more broadly "development" will be genuinely sustainable in longer-term and relatively less harmful ways. Therefore, my goal in this work is to describe this situation in a way that will be useful for those far better placed than I to act in ways that will have material benefit.

I therefore assemble Kedarnath-related data into a general picture in order to give a particular sense of the whole place that I hope will serve these goals. The design of the book aims to allow the reader to feel the argument and acquire a sense of the place and its longer story in a way that showcases the utility of an approach grounded in religious studies for understanding lived, emplaced human situations, particularly in the Himalaya (on this point, see Drew and Gurung 2014). I hope this book will also serve as an aid to people who work with questions of ecology, climate change, development, and tourism and who require a nuanced understanding of how cultural and religious contexts bear on such matters.

BHUPENDRA

It is, however, impossible to proceed further without introducing Bhupendra Singh Pushpwan, my friend and research associate. His notes, thoughts, questions, and insights infuse this book. He was in Kedarnath when the floods came and did not survive. I knew Bhupendra first as a friendly local, then as a schoolteacher, then as a research assistant and guide, and most importantly as a friend. Somewhere along the way I realized that he made a practice of befriending foreign tourists and becoming their guide, one of several Ukhimath men who do so, a logical and inescapable framework of relationship that would inevitably structure much of our encounter.

Bhupendra often had to patiently explain to me basic things about Garhwali culture and the Kedarnath valley and to tolerate my numerous idiosyncrasies and physical infirmities. I remember once, above Kedarnath when we went to gather Brahma lotuses to offer in the Kedarnath temple and were therefore required to walk barefoot, I walked so slowly that he took naps while waiting for me to catch up. I often had to reiterate my research aims because they were different from what he initially understood the object of my research to be. In one of our first official research meetings together, he offered me a series of quotations from the *Bhagavad Gita* that he had personally selected and that he thought was what I was looking for; I wanted to talk to him about how to start meeting and interviewing Kedarnath pilgrimage priests. My research allowance for a month was more money than he would make in a year. I was a white American working on a doctoral degree; he was an Indian *pahari* with what was called, in Ukhimath, ten plus two, two years of study in intercollege after ten years in primary and secondary school. And yet at the same time we found that we had much in common. We both thought of ourselves as late bloomers in life.

After my first several months in Ukhimath, I spent most of a year working closely with Bhupendra. We lived, cooked, and ate in the same room in Kedarnath during the 2007 season. He taught me how to cook dal, vegetables, and roti. He took me, reprising his sometime role as mountain guide, to the many villages and shrines of the Kedarnath valley that we agreed were important for my work. We once climbed up behind Kedarnath to see Chorabari Tal (also known as Gandhi Sarovar), the snowmelt-fed lake that would fall on Kedarnath in 2013. We carved "Luke and Bhupi" into a rock that probably later fell on Kedarnath. I went over with him, more than anyone else, how to phrase my questions, and often he would add his own. He advised me on when to participate, when to lie low, when to push. He had a gift for intercultural communication and clarity of expression; his background as a teacher served him well in his work with me. I think that working with me gave him the chance to exercise his considerable intellectual gifts much more than in any other job he had found so far in his life. I have belatedly realized that the clearest formulation of what I have come to understand as the eco-social complex power of Kedarnath came from Bhupendra when, during a walk, he turned to me and said, "Nature is the face of God." Many of my relationships with residents of the Kedarnath valley began through Bhupendra rather than through my own efforts or merits. His willingness to vouch for my worth paved the way for much of my work. After working with me, he went on to secure a job with an international nongovernmental organization and then later found his way back to Kedarnath working for a helicopter company, which is why he was in Kedarnath when the floods came. He leaves behind his parents, a younger brother, a wife, two children, and a village full of friends and family who miss him deeply. I hope that in some way this book will publicly illustrate what those close to Bhupendra already knew but perhaps he never fully managed to show the world: that he was an extraordinarily wise, kind, perceptive, and honorable person.

THE POWER OF KEDARNATH

Kedarnath is a place full of power. I use the term *power* primarily as a rough translation of the term *shakti. Shakti,* in both Hindi and Sanskrit, means "force," "power," "energy," and "authority." *Shakti* can also refer to one form of Mahadevi, whom some scholars term the "Great Goddess" of Hindu traditions. It is also a term that can be used to describe physical and social power and authority. Kathleen Erndl (1993, 32) has translated the *shakti* of Mahadevi as the "dynamic force inherent in matter." I trace the beginnings of my understandings of the *shakti* of Kedarnath to a rough memory of a conversation at the top end of the Kedarnath bazaar with a Kedarnath pilgrimage priest whom I will call Bagwadi-Ji, sitting against the wall of a lodge fronted by shops selling materials for ritual worship (Hindi: *puja*) along with pictures and books about the Uttarakhand Char Dham Yatra.[7] We were looking at the ebb and flow of people in the bazaar and in the temple courtyard just up the steps. Some were making reverences when they approached the temple. Some were snapping photographs and looking at the mountains. Some were tired and determined sedan-chair bearers who were carrying their charges all the way to the door of the agreed-upon destination. Some were seated ascetics (Hindi: *sadhu*) offering blessings and asking for donations as they leveraged their status as incipient Shivas. There is something here, Bagwadi-Ji said (my paraphrase), that makes people feel attracted (Hindi: *akarshit*) to the place. The "here" of Kedarnath pulls people. The noted Garwhali historian Shivaprasad Dabaral, citing Yashpal Jain's description of another Himalayan pilgrimage place (Amarnath), uses the same notion of attraction (Hindi: *akarshan*) to describe the power of the Himalaya in general.[8] Unlike the pilgrimage place of Badrinath to the west, which lies on a trade route crossing into Tibet, Kedarnath is on the way to no other destination. With the exception of herders who might have brought their flocks into the end-valley to graze, there is no reason for being in Kedarnath in at least the last one thousand years that has not directly or indirectly centered on Shiva and/or the reverence accorded the source of a tributary of the Ganga (the Mandakini River). The economic benefits of being in Kedarnath are predicated on the power of the place as a magnet of persons.

Bagwadi-Ji was, I felt, speaking to me on several levels. Conscious of his local knowledge of the site, he was offering a distillation of his professional understanding of the site "on the record." But he was also deeply embedded in his own assessment. Many of the people at whom we were both looking were potential clients who might commission him to perform ritual worship (Hindi: *puja*) on their behalf. Part of the attraction of Kedarnath that he was describing is economic. He would also have surmised that those visitors who were snapping photographs of the scene were less likely to feel a strong need for his services. Like many other *tirth purohits,* he was aware that nature-oriented tourism is increasingly important to the growing numbers of visitors to Uttarakhand each year and knew that many people did not come to Kedarnath with the primary goal of worshipping Shiva using traditional forms of ritual practice. Thus what he was noting was that Kedarnath attracts many people each year for different reasons but that all these people inexorably find themselves in the same spot.

The magnetic power of Kedarnath creates a feeling of vulnerability, of being a small creature pulled to a place that is not primarily set up for human comfort or even survival, no matter the amount or lack of available luxury. But this sense of vulnerability can take many forms. It can mean, for those who come to Kedarnath for the purposes of livelihood, economic vulnerability—no matter how much one can earn in a single day during the high season, there is no guaranteed income. It can be medical—even hardy Garhwalis who live in Kedarnath for any length of time struggle with illness and the effects of altitude. This sense of vulnerability is, since 2013, more easily seen. But it was there before as well. There is a precariousness to being a living human and, for locals, trying to make a living at almost 3,600 meters. To stand in Kedarnath is, among other things, to feel constrained by the power of what is there, a power that is mixed up with the presence of Shiva as well as the location itself. The fragility of human presence in the mountains is a persistent fragility because mountains are always changing. The weather is always changing. The trip up the steep, narrow valley functions as a focusing mechanism for this feeling.

In thinking about the power of Kedarnath I am also reminded of a conversation about the visual experience of God's true form (Hindi: *sakshat darshan*) from the Kedarnath pilgrimage season of 2007 that took place during a conversation between a family of *yatris* from Delhi and several local men. One of the local men, whom I will call Shukla-Ji, had spent much of his adult life working in Kedarnath during the pilgrimage season. While we had been around each other many times before, it was during this conversation that Shukla-Ji decided to offer his views on Kedarnath in my presence. The following is a paraphrase of what he said. First he told the story of a man who had come to Kedarnath, entered the temple, and, upon beholding the rock *linga* found in the inner sanctum, experienced *sakshat darshan*. For him this meant that he saw an anthropomorphic, matted-hair, trident-bearing Shiva in the rock. He exited the temple and went back two more times and experienced *sakshat darshan* each time. However, when he went again with his family it did not happen. A woman from the *yatri* family offered this interpretation: "The blessing God shares with you [Hindi: *prasad*] is according to your faith [Hindi: *shraddha*]."[9] I have often heard varieties of this sentiment expressed to me in Indian contexts when I am asking questions about divinity and experience. What a *yatri* brings to a place like Kedarnath determines how she or he will experience the encounter; the same could be said for a researcher. But this was not the end of the matter. As the conversation continued, the group agreed that there was something objectively distinctive about Kedarnath when compared to other famous Himalayan Hindu pilgrimage sites such as Vaishno Devi in Jammu and Kashmir. That is to say, place and divinity were not only in the eye of the beholder; the place and the deity had a say in the matter.

Shukla-Ji, normally a dignified and somewhat taciturn man, expressed this complexity with a passion indicating that this was something he had been thinking about for a long time. He said (my paraphrase) that he had been coming to Kedarnath for thirty-one years, had no reason to lie, and had never experienced *sakshat darshan*. On the one hand there was some power (Hindi: *shakti*) in the place. On the other hand, anyone could be God (Hindi: *Bhagwan*). I could be God. As he put it, "Heart and temple are places where people meet. People come here from all over the country and all over the world. They interact, speak with each other, and take the experience of being here [Hindi: *yahan ka anubhav*] with them when they leave."[10] For Shukla-Ji, much of what marked Kedarnath as a special place had to do with the distinctive social interactions characteristic of a famous Hindu pilgrimage place that attracted diverse populations for diverse reasons. What exactly constituted "the experience of being here" (Hindi: *yahan ka anubhav*) varied, but God and *shakti* were both *there* somewhere. The experience of meeting could be internal, external, and both. Yet at the same time, hovering quietly in the conversation, was the feeling that some power (Hindi: *shakti*) connected to Kedarnath operated independently of what humans brought to the encounter. This is also generally felt to be the case about a Hindu *tirtha* or *dham*. Surinder Bhardwaj and James Lochtefeld (2004, 479) explain this understanding of *tirtha* as "the deeply rooted conviction that certain places are powerful in their own right."

The power that Kedarnath exercises, and that is the focus of human attention, does not consistently localize in any one spot in the end-valley where Kedarnath is located. It is found in the place as a whole—constituted through the fuzzy and dynamic interplay of person, temple, and environment. This interplay can be felt and seen as the presence of Shiva or the residue of his presence, or it can be attributed the broader power of the Himalayan environment, itself infused with the earthy and aquatic presence of the Goddess and her *shakti*, or both.[11] It is hyperpresent but also opaque. It is attractive but cannot be grasped. It frustrates and relativizes human effort in a way that will not go away. On a good day this power is quiescent, a passive "palpable energy" (to use Lochtefeld's [2010, 201] term) that purifies and attracts by the mere fact of its nature rather than in a volitional, personified fashion. In May of 2007 Bhupendra was speaking with an old woman from Jaipur who had come to Kedarnath for the second time, the first being in 1992. He asked her about the relationship between Shiva and the earth (Hindi: *bhumi*) in Kedarnath. Her answer: "Lord Shiva and Kedarnath are the same thing."[12]

This sentiment is most easily understood visually. In trying to understand the complicated emplacedness of Shiva in Kedarnath, I found print images of Kedarnath available for purchase in the Kedarnath market to be of great use. *Yatris* usually purchase these images after visiting the temple and just before departure from Kedarnath. Such print images serve as souvenirs and often are ultimately installed in a home or work shrine. Some images of Kedarnath depict the *linga* inside the temple and show nothing else. Others simply show the temple and its panoramic Himalayan backdrop. Many images use the techniques of montage and collage to blend depictions of the inside of the temple, the external setting of the temple, and figural representations of Shiva and his partner the goddess Parvati. Looking across the available selection of print images depicting Kedarnath, one sees that Shiva is understood to be present inside the temple as the *linga* but also to be part of the place itself.[13]

From a human standpoint, Shiva's emplaced power is not wholly compassionate. Shiva-in-Kedarnath is both hyperpresent, utterly involved in the place, and at the same time aloof and uninvolved. Kedarnath orbits this Shaiva (Shiva-oriented) paradox. I remember one day in the bazaar in 2007 when a pilgrimage priest (Hindi: *tirth purohit*) told me (my paraphrase) that Shiva makes a poor personal deity (Hindi: *ishtadevta*), that is, the form of God you choose for yourself as opposed to the form important for your family, community, or village. Bhairava or Hanuman is better, he said—they are the type of deity that gets the job done (Hindi: *kam-karne-wale*). Bholenath (Shiva), he said, is fickle. Who knows if he will notice you or not? And if he does, who knows if it will be beneficial. I have thought back to this comment since the floods.

WORKING WITH THE IDEA OF "PLACE"

I view experiences of Kedarnath's power as mutually constitutive encounters between person and location, and this is reflected by my use of the term *place*. I understand place as the overlap of culture, social location, personal history, gender, sense experience, natural environment, and economic position that emerges at a geographic location or locations over time through the interactions of the location with embodied persons. My understanding is influenced by Edward Casey's idea that "place is the most fundamental form of embodied experience" (Feld and Basso 1996, 9).[14] I find particularly useful Casey's (1996, 26–27) point that places are to be understood more as events, processes, or happenings than as large, static objects. Experiences of place, like communicative acts, are continually emergent. Patterns in what continually emerges can be noted, but they are always in production and dependent on context. Casey's understanding of place is in turn predicated on an understanding of the person as an embodied subject who, following Merleau-Ponty, experiences herself as already embodied, already emplaced, and already connected to other embodied persons. That is to say, experiences of place are founded upon an embodiment that is inherently intersubjective. This understanding of place foregrounds an embodied, sensuous experience of being in the world. It is therefore an understanding both of *place* and of *experience.* Casey also notes, importantly, that "places gather . . . various animate and inaminate entities" (24). This sense of what Casey terms "gathering" will be important to remember.

HOW EXPERIENCES OF PLACE WORK

My understanding of experience differs somewhat from Casey's in that I think that socially, politically, and economically constituted forms of (inter)subjectivity are felt in much the same way as sense experience. This understanding builds on recent developments in phenomenological anthropology (Desjarlais and Throop 2011) or what some anthropologists term critical phenomenology (Willen 2007) or an "experience-near" analytic framework (Seeman 2009, 3–4; Wikan 1991). This approach aims to bring to articulation the experience of being in the world as it is framed by factors such as poverty, disease, illegality, trauma, and situations in which humans find themselves at the "limits of language" (Jackson 2008, xii; Seeman 2009, 3–4). A related approach is that of sensory anthropology, which has begun to think through the ways that the social is sensorially produced or understood as a kind of atmosphere (Chau 2008; Desjarlais 2003; Edensor 2007; Howes and Classen 2013).

Several related trends in geography have been pushing in a similar direction. One recent trend in critical geography, based on the work of Henri Lefebvre, looks at the ways in which the flows of money and political power through space create "spatio-temporal patterns," or rhythms (Edensor 2010) that are in important ways constitutive of human embodied experience. Others come to this way of thinking through the lens of *affect.* Nigel Thrift (2004, 64), for example, understands this modality of embodied geographic experience as "a sense of push in the world"—a push that "in the case of embodied knowledge . . . is provided by the expressive armoury of the human body" (see also Thien 2005). In this understanding, called by some "postphenomenological," of how places are involved with affect, landscapes are understood as the meeting point between the inner and the outer (Rose and Wylie 2006, 478; Rose 2006; Whatmore 2006). Some voices in this conversation theorize affect as something produced through processes in which humans are merely part of larger events that include a range of animate and inanimate entities (see, e.g., Lorimer 2008); this calls to mind Casey's sense of gathering. Scholars looking at place in South Asian contexts have begun to harness these theoretical tools as well (see, e.g., Sarbadhikary 2015).

Thinking about the feel of the power of Kedarnath with reference to these ideas of *experience, place, affect,* and *shakti* allows us to see how the power of Kedarnath is produced by the concatenation of many different factors and how deeply it registers in the embodied experience of people who find themselves standing in Kedarnath. It allows us to see how the political economies of mountain pilgrimage tourism in North India and the broader regional, national, and global forces that infuse the worlds of North Indian mountain pilgrimage tourism combine and fuse in the felt experience of those standing in Kedarnath, with the sensuous experiences of high-altitude travel, Himalayan weather, and divine *shakti*.

EXPERIENCING ECO-SOCIAL COMPLEXITY

But how do these different experience factors, or variables, relate to one another? I think we can best understand the way that experiences of place emerge by thinking of experiences of Kedarnath as experiences of an eco-social system that demonstrates fuzzy boundaries and is characterized by complexity.[15] The primary boundaries of the system are the end-valley, starting just after *devdarshini* and ending with the snout of the Chorabari glacier behind the village. The fuzziness of these boundaries can be illustrated linguistically. The term *Kedarnath* usually means only the village, or the temple. In the worlds of Puranic and Upapuranic Sanskrit texts, however, the words *kedara* and *kedaramandala* sometimes refer to the entire area of Garhwal and sometimes more specifically to the Kedarnath locale. There were temples to Shiva in his Kedareshvara form throughout the Indian subcontinent and even in Southeast Asia by the end of the first millennium CE. Heuristically limiting the primary field of analysis to the Kedarnath end-valley makes it possible to see patterns in the power of the place that would be lost at larger or smaller scales.[16]

The idea of eco-social complexity can be understood as the idea that humans and other elements of the world-organism each function as multiscale systems that exist in dynamic and nonlinear relation to one another (for an overview of this set of ideas, see Haila and Dyke 2006, 279–301). Embodied persons, water, rock, weather, ice, mud, horses, helicopters, buildings, money, social authority, political power, cement, stories, rituals, emotions, affects, Shiva, and *shakti* are all elements in the system and, if we refocused the unit of analysis differently, could be themselves understood as systems. Kedarnath is a complex system because there is no single predictable shape to how these different elements relate to and combine with one another. Rather, their relationship to one another is nonlinear. Environmental philosophers, environmental scientists, environmental historians, and environmental anthropologists use the idea of eco-social complexity to think about the patterns that connect the "sociocultural dynamics" found in human social systems with the complex systems of weather and climate (Haila and Dyke 2006, 17). Thinking about Kedarnath as a complex system also further underscores the necessity for broad sociohistorical contextualization. Complex systems are often characterized by their "extreme sensitivity to initial conditions," which means that the earlier history of how Shiva's presence in Kedarnath (and in the Himalaya more broadly) has been understood is in my view critical for understanding the shape of things closer to the present (Damon and Mosko 2005, 10). The idea of eco-social complexity also aids in conceptualizing Kedarnath-place-deity as agentive. Complex agency may be understood as the application of the idea of complexity to the notion of agency. It is the demonstration of coherent, dynamic, and "persistent patterns" of agency or action by a system of smaller elements or systems that appear to act in concert as part of a dynamic network (such as an economic system), a number of smaller entities or systems of different scales that overlap in a connected way (F. Cho and Squier 2013, 367–68).[17]

The complex system of Kedarnath is a system that can be characterized as an eco-social system. The idea of eco-sociality is an important tool for those who aim to think about the world in a way that does not place the human individual and human society at the center and regard the rest of the world as somehow set apart or separate from the human world. For example, Eduardo Kohn (2013) understands the relationship between humans and forests as relations between self-organizing and emergent systems that overlap and take account of each other. Tim Ingold (2000, 189–218) thinks of landscape as an ongoing, embodied, material dialogue between human and environmental organisms. Viewed in this way, disasters remind humans that their relationship to the natural world is an ongoing part-whole relationship. The routine occurrence of earthquakes, landslides, and floods that frame life in the Himalaya is a forceful reminder that eco-sociality is a basic feature of human existence. This understanding of eco-sociality also adds to the phenomenological idea of intersubjectivity mentioned earlier by noting that the intersubjectivity that is at the root of being in the world includes many different kinds of subjects, of which humans are but one subset.

My overall approach, the idea that we should think about experiences of the power of Kedarnath as experiences of a place that functions as a complex eco-social system, should be understood as a holistic, ecological approach. This way of thinking about Kedarnath is meant to showcase the potential offered by the discipline of religious studies for approaching global questions of critical importance to human life in the twenty-first century. Religious studies (ironically, given the European and more broadly Western history of the discipline) is useful not because I claim that it is itself a holistic discipline but because much of the data that it considers contains models of embodied personhood that connect to holistic worldviews, worldviews that frame human life as one part of a larger system that inflects across "differing scales" (Handelman and Lindquist 2011, 21). I rely upon the understanding of ecology fashioned by John Grim and Mary Evelyn Tucker (2014, 62) as part of their decades-long project to create a theoretical space for the study of ecology and religion: that which "locates humans within the horizon of emergent, interdependent life and does not view humanity as the vanguard of evolution, nor as the exclusive fabricator of technology, nor as a species apart from nature." For Grim and Tucker this commitment entails attending to how cosmological understandings embed in daily life. This kind of holistic treatment avoids offering further support to the deeply problematic and globally influential modern Western nature-culture binary, a binary that scholars such as Mabel Gergan, Ritodhi Chakraborti, Mona Bhan, and Andrew M. Bauer have argued continues to persist in ways of thinking about "the Anthropocene."[18] In thinking in these holistic ways I am also following the direction suggested by Manuel Vásquez (2011, 204–5), who has argued that a particular kind of holistic approach, what he terms (following Evan Thompson and Francisco Varela) "relational holism," might serve as a theoretical charter for religious studies.

The idea that the religious dimensions of the situation are somehow separable from the situation of the place as a whole is another intellectual predisposition this book seeks to undo. In one sense it could be said that in this book I use the analytic category of place to examine the overlap of political, economic, cultural, social, religious, and environmental factors in the lived experience of Kedarnath. Ultimately, however, the goal of doing things in this way is to provoke the realization that this idea of "overlap" is based on a false premise—that there are separate categories that overlap. Conversations about sustainable development, colonialisms and postcolonialities, climate change, political economy, religion, spirituality, nationalism, and modernity still appear to many to be putatively separate yet interconnected phenomena. We are moving past this assumption of categorical, a priori difference, with those who study geography and "environment" across the humanities, social sciences, and sciences leading the way. But there are still taxonomies to unthink, and ironically, given the Western origins of the concept of religion, the doing of ecology and religion can help in this regard. It is rare that scholars of tourism or Himalayan development seriously engage, for example, with the technical details of Shaiva (Shiva-oriented) philosophy. David Haberman (2013) laments an analogous kind of taxonomical myopia when he discusses how Western scholars, for the most part, were unable to comprehend the centrality of tree worship to South Asian religious worldviews because they were unable to see trees as potentially living beings in a way close to how humans are living beings. Should this blindness be cured, Haberman argues, Western ways of thinking might eventually take a more inclusive path. In the case of phenomena such as pilgrimage tourism, Himalayan development, and climate change in the twenty-first century, a commitment to start at the overlap might ultimately create the conditions for better long-term planning and a greater willingness to act with foresight in ways that keep religion, ecology, and development in the same conversation.

SACRED PLACES

Recent work in the study of religion on pilgrimage, space, and place has concerned itself with testing and critiquing the utility of two famous theoretical approaches that shaped a great deal of scholarly work in the late twentieth century. The first is the contention of Mircea Eliade that sacred power can be generated directly by or is intrinsically present in a physical location. The second is the contention of Victor and Edith Turner that journeys to places of religious significance (and by extension the places themselves) occasion a collective feeling of *communitas,* a powerful and transformative temporary erasure and inversion of normal social relations.

Several different strategies for engaging these approaches in some way or other acknowledge that certain physical locations in the world demonstrate a distinctive power of some sort, a power characterized by the often-ritualized journey of people to that location. The first is to preserve the use of the term *sacred* and to return to the Durkheimian roots of the idea of the sacred, regarding it as a form of power primarily constituted by social forces. As Jacob Kinnard (2014, 2) puts it: "It seems almost self-evident to say that religiously charged places and spaces, as well as the images associated with them, become powerful because they are made powerful. That is to say there is nothing inherently sacred about any place or space or physical object; human agents give them power and maintain that power." In the Durkheimian sense, "things sacred" are collectively designated—they are set aside from profane things through mechanisms of community and society. The Turnerian idea of *communitas* similarly posits a certain kind of social unity, albeit a temporary one. In recent decades this presumption of social unity has been challenged and to some degree replaced by a model that regards social contestation and conflict as the most salient social factors (Coleman 2002; Coleman and Eade 2004; Chidester and Linenthal 1995).

Another approach is to reject the utility of the term *sacred.* There are several linked reasons for this rejection. One is that the Eliadean echoes of the term imply the independent existence of "the sacred" as a transcendent entity, something that some recent conversations in the study of religion now question. Another reason is that it is not a term of any heuristic utility because, as a function of its post-Enlightenment Western origin, it presumes a strict separation of sacred/profane or sacred/secular that either is not present in the example under consideration or is produced in ways that are so robustly contingent and fluid that it does not make sense, even heuristically, to posit the binary. For example, Kama Maclean (2008, 9), in her work on the recent origins of the Kumbha Mela, asserts that "the assumption of distinct spheres of secular and sacred activity has obfuscated a clear understanding of pilgrimages as powerful material forces." In the case of pilgrimage in Japan, Ian Reader (2014, 15) has argued that that the "dynamics of the marketplace . . . are not antithetical to pilgrimage (or to 'religion'), but crucial to its successful functioning, development, appeal, and nature." Sarah Thal (2005), using examples drawn from Japanese contexts, has demonstrated that particular constellations of social, political, and economic forces can over centuries transform the character of a particular pilgrimage place almost beyond recognition.

Thus of paramount importance today in the study of religion is the recognition that places and spaces of religious significance are mutable, human products that wear diverse and historically contingent faces. Today many studies of locations possessing sociocultural, "religious" significance emphasize the diversity and multiplicity of how people experience such places. In his recent and definitive study of the Hindu pilgrimage place of Haridwar, just several hundred kilometers to the south of Kedarnath, James Lochtefeld (2010, 5) argues strongly for the idea that the city can be understood only through "multiple narratives." In their recent edited volume on Varanasi, Martin Gaenszle and Jorg Gengnagel (2006, 8) succinctly summarize a welter of anthropological literature on this point: "The rich literature of an anthropology of space and place has shown that there is great variety in the ways 'sites' are culturally constructed, and that one and the same place may be seen rather differently by different actors."

However, the emphasis on the constructed, multiple, and changeable character of place and pilgrimage, by design, does not aim to fully register the depth and material, experiential weight of what some people find some of the time when they travel to or reside in such places. Jacob Kinnard (2014, xv–xvi) is, I think, attentive to this issue. He bases his exploration of the religiosity of places in the modern world in the models of Henri Lefebvre and Edward Soja that posit three interwoven levels of spatial experience: the materiality of location (Firstspace), "imagined space" (Secondspace), and lived realities of the everyday where these two "intertwine" (Thirdspace). This model is a usefully flexible way of approaching many places of religious significance, particularly if, like Kinnard, one is trying to look across a number of examples that are very different from one another (in Kinnard's case Tarapith, Ground Zero, Bodh Gaya, Karbala, and Devils Tower National Monument). My approach follows Kinnard but places, relatively speaking, greater weight on the felt experience of place.

THE ECO-SOCIALITY OF SACRED PLACES

Recent decades have seen scholars across the humanities and social sciences using the ideas of flow, system, and network to theorize, in many different ways, the nuances of how embodied persons experience being in a world constituted by the material weave of other persons; objects; biological, social, political, and economic processes; relations of power; symbols; ideas; life forms; and environments. These ideas are now part of the always-changing toolbox of religious studies.[19] One prominent voice in these conversations, Manuel Vásquez (2011, 319), has specifically urged scholars in the study of religion to draw on the study of ecology in order to better grasp the nuances and textures of these complex systems of material, embodied experience. This way of thinking about religious experience functionally rejects the idea of the sacred as a separate and immutable feature of the world and instead shows how the materiality of the world assembles in concert with the cultural and the biological in ways that become religiously powerful. I understand these approaches as attempts to account for religious power and experience that are simultaneously nonreductive and nonessentialist. Approaching Kedarnath as a complex eco-social system whose complex, eco-social character emerges in conversation with embodied persons thinks in concord with these trajectories but in a way that focuses with urgency on a particular place and emphasizes the part-whole relationship of humans to the world in which we are resident. The protean power of Kedarnath arises out of the conversation of environment, society, culture, deity, and person and does not reduce to any single feature of the place. My theoretical synthesis also attempts to bridge some of the analytic distance between these approaches and the cosmology-based framework of Grim and Tucker by attending closely to the material, political, economic—contingent—aspects of exactly how cosmologies are embedded in the everyday. In this regard I follow the direction mapped in in the recent edited volume Nature, Science, and Religion: Intersections Shaping Society and the Environment (Tucker 2012). Politics, money, weather, environmental history, theories of sustainable development, the dynamic interplay of Shiva and *shakti*—all come together, in highly contingent ways, in the experience of people who find themselves standing in the marshy ground of Kedar.

THE STRUCTURE OF THIS BOOK

The persistent pattern of Kedarnath-as-place is the primary character in the plot of this book. After 2013, the life story of this persistent pattern of place now divides into a "before" and "after." I want to both honor and resist this division. I follow the "persistent pattern" from its historical beginnings in the second part of the first millennium CE into the *apda* (disaster) and through the beginnings of the post-2013 tale of the place in a way that serves as an act of testimony about what was lost, interrogates the preconditions of the disaster, and interprets the beginning post-2013 stages of the continuing complex processes of this place. That is the story of this book—how increased pilgrimage tourism, unsustainable development, and a twenty-first-century monsoon collided with a famous *dham* (Hindi: abode) of

Shiva in the Himalaya, why the collision had the force that it did, and what can be learned from the collision.

Chapter 1, "In Pursuit of Shiva," uses the story of the pursuit of Shiva by the Pandavas to introduce the Kedarnath of recent times. I introduce the social worlds of Kedarnath and discuss how I fit in to those worlds. I argue that the purificatory power of the place and the character of the god resident in the place call into question the necessity of approaching the place with virtuous conduct and purity of intention, a narrative framing that surrounds the present. Chapter 2, "Lord of Kedar," examines the history of Shaivism in Uttarakhand and treatments of Kedarnath in Puranic and tantric texts. Here I argue that the shape of the Kedarnath *linga* indexes an archaic and specifically Himalayan understanding of Shiva's enmeshing with the world that has been formative in creating the "persistent pattern" of the place as a whole. Chapter 3, "Earlier Times," examines how Kedarnath transformed from a remote destination for renouncers and kings into a popular destination for pilgrimage tourism located in a state famous for its environmental activism. The geographic remoteness of the *dham* intensified the impact of the changes brought on by new statehood and a massive rise in the number of *yatris* coming to Kedarnath in ways that reflect the turbulent environmental history of the region.

Chapter 4, "The Season," describes what Kedarnath looked like during my fieldwork in 2007 and 2008. I show how the place was bursting at the seams and note that many of the traditional modalities for visiting and living in the *tirtha* had begun to drop away. In 2013 the waters descended, and chapter 5, "When the Floods Came," tries to tell the story of those early days and weeks surrounding the floods in June 2013. Then, in chapter 6, "Nature's *Tandava* Dance," I unpack early reactions to the disaster to demonstrate how new factors (development and tourism) joined and changed the connotative and experiential web connecting Shiva and *prakriti* (Hindi: nature, also a form of the Goddess in some contexts) in Kedarnath. Chapter 7, "Topographies of Reinvention," concludes by reviewing the present situation and reflecting on the broader lessons that can be learned from a close examination of Kedarnath.

1

In Pursuit of Shiva

"This place is not meant for people." Men who lived and worked in Kedarnath would often express this sentiment to me. I would remember these words at odd moments when, during my time in Kedarnath, the remoteness of the place would suddenly strike me. These moments often happened when I was waiting in line to enter the temple. In the years just before the floods, it was sometimes necessary in the morning to wait for several hours in line for the chance to enter the temple and come face to face with the self-manifest rock form of Shiva around which the temple had somehow been built out of massive stone blocks high up in the Himalaya. This time spent waiting in line was many things: a chance to talk to *yatris*, the opportunity to observe the spectacle of pilgrimage tourism from a ringside seat, an embodied push for me and my cold feet to think about why I had come to this place and what I was trying to do.

Once while in line I remember looking into one of the shops that surrounded the temple on three sides and seeing that a film was playing on a small television. On the screen, hanging above a counter full of metal trays containing materials (Hindi: *prasad*) that would be used in ritual worship (Hindi: *puja*) in the temple, next to a few benches where one could sit and take chai and biscuits, I saw the Pandava princes and their wife Draupadi making their way up into the Himalaya, struggling along on their famous ascent to heaven. The scene reminded me that one of the most famous stories about Kedarnath connects to narrative material that in Sanskrit versions of the *Mahabharata* epic is found in the Chapter of the Great Departure (Sanskrit: *Mahaprasthanika Parva*) and the chapter of the Climb to Heaven (Sanskrit: *Svargarohana*). In many ways those who find themselves in Kedarnath are walking in the footsteps of the Pandavas. Kedarnath is bound up with a story about trying to leave the world behind, a story that in many versions includes eventually walking out of the living world on what is known as the Great Path (Sanskrit: *Mahapatha*; Hindi: *Mahapath/Mahapanth*). Yet at the same time it is a place that forces humans to reckon with their place in the broader world, where the eco-sociality of human life becomes obvious. It is a transition zone. This journey of the Pandavas up into the Himalaya is also, more broadly, one of the most important stories connected to the Land of the Gods (Hindi: Dev Bhumi), a recent popular designation for the Garhwal region of Uttarakhand. This label has, in recent years, come to serve as a brand identity for the region, although like many features of the geographic imaginaire that connect to stories of Hindu deities and their actions in the world, it sounds unique to the region, yet is in fact not unique.[1] The neighboring state of Himachal Pradesh has found its way to the same designation.[2]

Accounts of the Great Departure of the Pandavas offer important but confusing lessons about how humans should act in the world—they provide a narrative framework for thinking through ethical commitments, and in this way these stories function as a microcosm of the *Mahabharata* story in general. While living in Kedarnath, I came to realize that a careful study of the dynamics of the line to enter the temple touched on similarly complicated and crucial issues, namely the relationship between the purity of one's aims and the nature of the end achieved. Thus, as a way of introducing the social and spatio-temporal worlds of Kedarnath, I want to show how narratives about the arrival of the Pandavas in Kedarnath help us see how the social, moral, and theological issues raised by the dynamics of the line (or queue) can serve as a general introduction to Kedarnath as it was in the several years before the disaster. In this chapter I will therefore first briefly describe Kedarnath village, relate and contextualize an oral version of one of the most famous foundation stories of Kedarnath that involves the Pandavas, and then discuss how waiting in line in Kedarnath analytically functions as a window into both the place as a whole and my position in that place. As I came to understand, to have any sort of involvement with Kedarnath in which the stories of the Pandavas figured prominently was to stand in a place framed by the closeness of the end of life, the hyperpresence of an aloof god, and the challenges built into the human pursuit of virtue.

THE SHAPE OF THE PLACE

When I stood in line for the temple in Kedarnath, I, like most visitors, looked out over the village and remembered the journey that had brought me to this mo-ment in space and time. I will use my memories of this view as a springboard for a brief introduction to the place of Kedarnath. This description, like the opening *devdarshini* vignette, remembers the scene as it was before 2013. The layout of the village is different today. The path from Gaurikund to Kedarnath used to wend its way up the western bank of the Mandakini River valley to the end-valley where Kedarnath is located. The river valley was formed by several advances and subsequent retreats of the Chorabari glacier, a mass of ice and rock that lies today just behind Kedarnath village, between the village and the high Himalayan mountains just beyond to the north (Chaujar 2009; Mehta et al. 2012). With every kilometer of the approach to Kedarnath it was possible to feel the imprint of the glacier. The tree line passed, the lines of the surrounding valley walls began to feel just slightly aquatic and the bare bones of the mountains emerged. Usually at the beginning of the season (which often falls in May) the winter snow had not completely melted, and the second seven of the fourteen kilometers passed through snowbanks. At *devdarshini* the ascent lessened and the narrow river valley opened out into a wide, open split-level moraine framed by the glacier behind and mountains on either side. Diana Eck (2012, 228–29) describes the journey in this way:

> Pilgrims strike out early in the morning to walk the steep trail over ten miles to Kedāra. Finding their own pace, they stretch out along the trail—ponies in the lead, followed by pedestrian pilgrims both with and without shoes, and four-man *dhoolies* carrying the elderly or infirm. There are white-clad widows in tennis shoes, their walking sticks an absolute necessity, some making their way slowly with one stick in each hand. There are Rajasthani women in bright skirts and thick silver anklets, walking barefoot. They are robust old men wearing jackets and woolen mufflers, with porters carrying their gear; and there are scantily clad *sannyāsīs* carrying nothing but a water pot and a bowl. Parties of pilgrims from Gujarat and Karnataka mingle with families from Maharashtra and schoolgirls from Bengal. They cross high snowfields and marvel at the snow. For most, it is the first snow they have ever seen.

Before Kedarnath proper I used to arrive into what was called the horse camp (Hindi: *ghora parav*), a milling confusion of riders attempting to mount and dismount, passengers entering and exiting sedan chairs, mud and horse urine, chai stalls, and, for a month in 2007, a temporary cyber-cafe. If you needed to shave your head in order to observe *shraddh* (Hindi: the rituals of ancestor worship), it was here that you would have had a barber do it before you entered officially into Kedarnath. Directly parallel to these last two kilometers on the eastern bank of the Mandakini was the helicopter area with two landing pads shared, in 2008, among at least six different helicopter companies. Yatris reaching Kedarnath on the path would look across the river to see *yatris* arriving by helicopter, for many the first time they were seeing a helicopter up close. Up on the edge of a side valley to the east in good weather one could see the rocky outcropping, flags, and tridents of the shrine to Bhukund Bhairavnath, the guardian deity of Kedarnath. To the west, on the slope of the valley side, a stream powered the electric generator that supplied the electricity for the buildings of the Badrinath-Kedarnath Temples Committee (hereafter referred to as "the Samiti," Hindi for "committee" or "association") and the government. The rest of the village was on a separate power grid that depended on electric wires that ran up from Gaurikund.

After the horse camp, the path crossed the Mandakini River on a sturdy metal bridge. Just after crossing, many, though not all, would take a moment to purify themselves with a bath at the bathing steps (Hindi: *ghat*) on the Kedarnath side of the freezing waters of the river. Since 2013, as part of a massive postflood reconstruction effort, the bathing *ghat* has been completely rebuilt and enlarged. Then the path forked, on the left turning up into a collection of lodges and *dharamshalas* (Hindi: pilgrim guesthouses), starting with Maharashtra Mandal. The main fork of the path to the right wound steeply up and curved around in an s-shape until the beginning of the main bazaar was reached and the Kali Kamli *dharamshala* lay on the right. At this point the temple could be seen straight ahead and higher up at the other end of the bazaar. I often stopped at this point for a moment. *Chai* stalls and *dhabas* (Hindi: open-air restaurants) were found here, but the majority of the real estate looking onto the bazaar was held by shops selling the material objects connected to Hindu pilgrimage in the Himalaya: warm clothing, snacks and staples, materials for *puja,* books and pamphlets about Kedarnath and many related and standard North Indian religious Hindu topics, prayer beads made from the seeds of the Eye-of Shiva tree (Hindi: *Rudraksh*), containers for water from the Mandakini, small *lingas,* plants useful for Ayurvedic treatments (particularly those of potent Himalayan provenance), and most visibly thousands of images of Kedarnath, Shiva, the Uttarakhand Char Dham, and other North Indian Hindu religious staples such as Vishnu, Krishna, and Durga. When I noticed the Ayurvedic plants I remembered the famous biodiversity of Himalayan flora. Sometimes I would recall the moment in the *Ramayana* when Hanuman flies to the Himalaya to find the magical life-restoring plants that will help to save the wounded Lakshman and, uncertain which plants to bring, carries back an entire mountain. About halfway through the bazaar on the western side one saw the building that had been built over the water tank of Udaka (Hindi: Udak Kund), whose water was often connected to a famous Sanskrit verse: "When one has drunk the water of Kedara, rebirth does not occur." Just beyond Udak Kund was a small Goddess temple. At the end of the bazaar a flight of steps led up to the raised temple courtyard, where the triangular mountain-shape of the temple forced itself up into the sky. If I was acting like a typical Kedarnath visitor, I would celebrate having reached my goal. I would ring the bells hanging over the entrance to the courtyard at the top of the stairs. Sometimes I would just "take *darshan*" (Hindi: *darshan lena,* the interactive visual encounter of devotee and deity) of the temple from a distance and then, fatigued, go to my room to rest for a time. At the beginning of the season in 2014 there was so much mud, a year after the flooding, that the stairs were still buried and the path through the mostly ruined bazaar led straight into the newly pristine temple courtyard.

Behind the temple to the northeast were, and are, the residential quarters of the Samiti, the *pujaris* (Hindi: ritual specialists tasked with the daily worship of temple deities) and the police, as well as two cell phone towers. To the northwest behind the temple were several more *dharamshalas,* most notably Bharatsevashram, a monument and adjacent hall dedicated to the famous philosopher Shankaracharya, and the free food kitchen (Hindi: *langar*) and ashram of the renunciant Mahant Chandragiri, adjacent to a railway reservation office that in 2007 was still under construction. Mahant Chandragiri's ashram was a place where free food and a place to sleep were always available whether you were a renunciant, a Russian Shiva devotee, or an American graduate student. Bharatsevashram, which was, and is, part of a national organization founded by Bengali guru Swami Pranavananda, was used primarily by Bengalis and tour groups from Gujarat. The line for the temple ran next to Bharatsevashram for much of its length. To the immediate east and northeast of the temple were the offices of the Samiti, where *yatris* could buy tickets for special *pujas* and seating inside the temple during evening worship offering light and songs to deities (Hindi: *arati*). Again moving eastward, the Samiti had built a large cloakroom for the storage of belongings and a public auditorium to host events such as, in 2007, a recitation of the Sanskrit text of the *Shiva Purana* and a concert by the famous Garhwali singer Narendra Singh Negi. Just by this auditorium lay the Swan Water Tank (Hindi: Hams Kund, a water reservoir linked to a story in which the deity Brahma assumed the form of a swan), the location in Kedarnath for *shraddh* rituals. At a somewhat further remove on the western and northwestern side of Kedarnath were the small glacial lakes of Vasuki Tal (up and to the west) and Gandhi Sarovar, or Chorabari Tal, as it was known locally (up and to the northwest). Many intrepid *yatris* used go to Gandhi Sarovar; one could walk there and back in half a day. A hardy few went to Vasuki Tal. One older route to Kedarnath, used by *yatris* coming on foot from Gangotri, proceeded from Triyugi Narayan to Kedarnath via Vasuki Tal. Some survivors of the flooding in 2013 escaped this way. "Brahma's Cave" (Hindi: *Brahma-gufa*) reputedly lies near Gandhi Sarovar. To the northeast of the temple reputedly lies the valley entrance to the Great Path, and near it, according to legend, stands the rock of Bhrigutirtha (also known as Bhairav Jhamp), from which devotees once jumped as a way of transitioning out of life. When the floods inundated Kedarnath in 2013, much of the water and debris came from Gandhi Sarovar.

When I would wait in line in the morning in 2007 in the high season, the line usually stretched back to the northwest of the temple, past the Shankaracharya Samadhi monument that commemorates the putative visit of one of India's most famous philosophical theologians and religious leaders, the philosopher Shankara. It extended past the Bharatsevashram *dharamshala,* the most popular destination

FIGURE 5. Waiting in line, with many one-use plastic raincoats attesting to recent rain. Temporary tents belonging to several renunciants can be seen to the left of the queue, with Mahant Chandragiri's free food kitchen directly behind and a cell tower off to the right.

in Kedarnath for the groups of Bengalis who visited, most often during the Durga Puja festival season in the fall. It curved past Mahant Chandragiri's *langar*. Even at its busiest, the line did not quite reach the two cell phone towers whose markedly well-constructed material modernity marked the northern edge of the village. To wait in this line, you would need to be ready for several hours of cold, numb feet.

Standing in line was a good way to encounter, in one visual sweep, almost all of the diverse social worlds that overlap in the Kedarnath of recent times. I saw groups of *yatris* arriving up through the market. Some had come by foot from the trailhead at Gaurikund, while others had ridden on a pony or been carried up in a sedan chair. Increasing numbers of *yatris* had been arriving by helicopter. Visitors in Kedarnath were usually Hindus from all walks of life who either were from Uttarakhand or had journeyed to the Land of the Gods from all over India and Nepal, though the majority were visitors from North India who were visiting Kedarnath as part of the Uttarakhand Char Dham Yatra. Increasingly *yatris* were arriving in Kedarnath ready to pay, in the high season, over one thousand rupees per night for a room, ten or fifteen rupees for a chai, and between fifty and one hundred rupees for a hot meal. The many who arrived in Kedarnath with fewer resources, such as a village group from Madhya Pradesh or Rajasthan on a once-in-a-lifetime *yatra*, had to make do with lesser options. Some were clearly in happy shock at finding themselves in such a place, while others were physically and mentally overwhelmed, ready to find where they would be staying and get oriented. Four-man porter teams carrying their passengers in palanquins would sometimes carry their charges all the way up the main road of the marketplace. Renunciants, most but not all of whom were men, sat at various points alongside the queue or in the temple courtyard or in and around the shops and smaller alleys that partially surrounded the temple. Some were permanent and semipermanent fixtures who came every year for part or all of the season. Others were passing through and would stay for a few days before moving on to their next temporary destination. Wearing few clothes, often smeared with ashes, they were the human icons of Shiva in his place. They received, and often demanded, donations in exchange for blessings.

Kedarnath *tirth purohits* (Hindi: pilgrimage priests, also known in Hindi as *panda* or the more formal but less used *tirth-guru*, pilgrimage guru) would often approach the line to offer their services as ritual specialists to those who had not already connected with their ancestral pilgrimage priest upon arrival in Kedarnath. These individuals have the hereditary job and right to host those who have come to a *tirtha* or *dham* on *yatra*, perform rituals on their behalf, and collect payment for those services. Traditionally, part of what a *tirth purohit* does is to offer his patron (Hindi: *yajman*) an introduction to the place that includes a description of how a particular god or goddess came to be present in the place and what the *yajman* needs to know about what should be offered in the temple or other place-specific ritual details, when to wake up in the morning, and when to depart in order to take best advantage of rapidly changing weather.

Kedarnath *tirth purohits* live in roughly three areas in the Kedarnath valley. The first and most socially central is the Bamsu area just south of Guptkashi, the area that traditionally belongs to a demon (Hindi: *rakshas*) devotee of Shiva named Banasur.[3] The second is from just north of Guptkashi up to Fata (from Nala to Fata on the map, including villages not on the map such as Khat and Rudrapur). Fata is the high-water mark for how far north into the Kedarnath valley the *tirth purohit* families reside. The third group of *tirth purohits* live on the other side of the Mandakini River in and around Ukhimath. Kedarnath *tirth purohits* count themselves as descended from an original group of 360 families, or the "three-sixty."[4] Semwal, Bagwari, Shukla, Sharma, Avasthi, Posti, Tiwari, Shastri, Lalmohariya, Jugran, Kapruwan, Purohit, Kotiyal, Vajpai (Vajpayee), and Trivedi are their main family names. Different families (and subdivisions within families) maintain relationships with specific groups of *yatris* (constituted by a combination of region, usually a North Indian region, and caste group). Particular *tirth purohit* families have historical relationships with, for example, specific Marwari families from Rajasthan, certain villages in the greater Allahabad area, a particular district in Rajasthan, or particular groups from Himachal or Jammu-Kashmir. However, since these are historical relationships, the same Marwari family might live in Mumbai or Calcutta, or those from the Alwar district in Rajasthan might reside in Patna or Pune.

Often renunciants would make their temporary homes near the line that formed for the temple and would ask *yatris* for donations as they approached the temple. A few more established renunciants would take up specific positions in the temple courtyard every day. James Lochtefeld has observed that more renunciants were found at Kedarnath than at other shrines such as Badrinath and that their presence at the site was in spatial and social terms more central.[5] Sister Nivedita (1928, 43–44) made a similar observation in her visit almost a century previously. Even when they behaved in ways that other Kedarnath residents and *yatris* found inappropriate (fighting, requesting donations too aggressively), renunciants were given an extreme degree of latitude in Kedarnath because they essentially appeared as forms of Shiva (garbed as anthropomorphic Shivas) in the place of Shiva. Both through observation and in fragmentary conversations over the course of the season, I realized that for many *sadhus* in Kedarnath the most important and foundational religious practice was simply staying in Kedarnath itself rather than spending a great deal of time in the temple. For the renunciants who found the bustle of Kedarnath a challenge for their goals of a quiet and inwardly focused residence, there was (and is) a Ramanandi ashram two kilometers south of Kedarnath, just below *devdarshini* at Garud Chatti.[6]

This ashram had a core community of several renunciants who even stayed in Garud Chatti through the winter, and it welcomed renunciants throughout the pilgrimage season as well.

Occasionally small groups of trekkers would arrive into the village, and sometimes I would be in line when they marched proudly past. Kedarnath has been a destination for trekkers of primarily South Asian (and often Bengali) origin whose sophisticated rucksacks, better footwear, and more weather-appropriate clothing mark them as distinctly different from other groups of visitors. I occasionally saw groups of Western trekkers, but nothing like the numbers who traveled to Gangotri and onward to the source of the Ganga in Gaumukh. The popularity of Uttarakhand as a destination for nature tourism has been a rapidly expanding segment of the pilgrimage tourism industry in recent decades. Skiing in Auli, rafting on the Ganga, walking in the Valley of Flowers, day trips from the hill station of Mussoorie—these have become well-known and popular activities in their own right that may be carried out as part of an Uttarakhand Char Dham Yatra or separately and can be found in Kedarnath as well. One of the primary missions of the Garhwal Regional Development Authority (Hindi: Garhwal Mandal Vikas Nigam or GMVN for short), which is part of the state tourism ministry, is to support these forms of pilgrimage and nature tourism by providing guides and appropriate lodging and support everywhere that sightseers might wish to go. In Kedarnath the GMVN was often where large groups of trekkers would stay because it was relatively easy to book the rooms in advance from outside the region and because the quality of the rooms and food closely matched their expectations.

As I inched closer to the temple, the line became a good vantage point from which to survey the commercial heart of Kedarnath—the main road of the bazaar. Some of these shops were owned by Kedarnath *tirth purohits*—not all Kedarnath *tirth purohits* who worked in Kedarnath performed *puja* for their patrons. There were also a number of businessmen who lived in Kedarnath during the season who were not *tirth purohits* and who did not perform *puja* for *yatris*. Many of the shopkeepers, merchants, restauranteurs, and lodge and *dharamshala* managers in Kedarnath were also from the Kedarnath valley but were not *tirth purohits*. They would have been from a different Garhwali Brahmin community or would have been from Rajput castes, known locally by the general umbrella term of *thakur*. Men from almost every village in the Kedarnath valley worked in and around Kedarnath during the pilgrimage season. That is one of the many reasons that the impact of the floods of 2013 was so terrible. Everyone had friends and relatives who were working in Kedarnath at the time. Even the most menial job connected to the pilgrimage industry paid better than most other available opportunities unless someone was employed by the government, the military, or the police. There was a definite cachet to working in Kedarnath itself. The tiniest retail space in the Kedarnath village could easily provide comfortable living for a year. Kedarnath pulled people and money from all over the nation and all over the world.

On those days when I arrived at the front of the line in participant-observer *yatri* mode, I touched the feet of the Ganesha who stood outside the temple door and, with the assistance of a *tirth purohit*, offered a quick preliminary *puja*. At the doorway of the temple I was greeted by an employee of the Samiti whose job it was to note down the number of people in each *yatri* group and record each group's point of origin. This benign encounter with recordkeeping signaled an important feature of how social space worked in Kedarnath. The Samiti officially controlled the temple. Inside, Samiti officers maintained order in the line, and men with the title of Veda-reciter (Hindi: *Ved-pathi*, the designation of those Brahmin Samiti employees who did *puja* in the Kedarnath temple) sat or stood by their designated posts inside the temple, ready to perform the same ritual services offered by the Kedarnath *tirth purohits* but without the same necessity of negotiating a ritual fee, instead accepting donations that were given directly into a donation box. The Samiti, at its discretion, allowed people to enter by the side door and jump the line. The only others with the right to jump the queue were, by consensus, Kedarnath valley locals who worked in Kedarnath. During my time in Kedarnath it was clear that the Samiti had struck deals with particular helicopter companies—on many occasions if you arrived by helicopter you were treated as a VIP and you could enter the temple without waiting in line. VIP guests of particular Kedarnath *tirth purohits* did not receive the same privilege, which meant that the question of side-door entry became the focal point through which the broader tensions between *tirth purohits* and the Samiti were frequently expressed. There was litigation between the two groups about the rights of *tirth purohits* to collect fees for their services in the temple courtyard and the Samiti's ostensible abrogation of those rights, a legal point that stood in for deeper issues about who controlled what in Kedarnath. Once through the door you entered the antechamber of the temple and began to walk around Shiva's bull vehicle Nandi in the auspicious direction, taking *darshan* as you did so of Draupadi, Arjuna, Lakshmi-Narayan, Kunti Mata (the mother of the five Pandava brothers), and Yudhishthira. Then you crossed another threshold into the hallway leading to the inner shrine where Kedarnath-Shiva sat, the god embedded in and arising out of the mountain rock. You stepped through a final threshold and were in the inner sanctum, in the presence of the god. After *darshan* and, often, *puja*, you briefly worshipped Parvati in the hallway before exiting back to the antechamber, where you completed the circumambulation of Nandi and took *darshan* of the remaining Pandava brothers and Krishna before exiting through the side door and circumambulating the temple.

The temple used to close in the early to midafternoon so that the inner sanctum could be cleaned of the material from morning *pujas* and so that Shiva could be offered food (Hindi: *bhog*) and could be worshipped with a fire ritual (Hindi: in this case *yagya havan*, or just *havan*, in which offerings are made into a fire while *mantras* from the Vedas and other Sanskrit verses are chanted). Once the temple closed for the afternoon, even the Garhwali deities who arrived on *yatra* with their villages, carried in palanquins or backpacks containing metal masks that served as the material embodiment of the traveling deity, had to wait. Such deity processions were a common and important feature of religious life in many Himalayan areas, most notably in Garhwal, the neighboring region of Kumaon, and the state of Himachal Pradesh. The deities traveled accompanied by their villagers, drums, and insignia (Garhwali: *nyauj-aur-nishan*) such as flags, and their arrival into the village was always an intense and exciting moment. After *bhog* the doors reopened in the late afternoon for *shringar arati* (the offering of fire, light, and music to the adorned [*shringar*] *linga*) and were worshipped with adornment and lighted oil lamps. At this time only the *pujari* and his assistant were allowed in the inner sanctum. A *pujari*'s job is to offer regular daily and festival worship to the deity/ deities of a temple regardless of whether devotees are present. It is not worship on the behalf of specific human devotees but rather worship that is an intrinsic part of the daily rhythms of the temple and that must be maintained as part of the ongoing human-divine relationships that center in a temple. The *pujaris* in Kedarnath were (and are) from the Shaiva denomination known today as Virashaiva and more specifically from the *jangama jati* class of a segment of the Virashaivas connected to the Five Teachers (Sanskrit: *panchacharya*) tradition.[7] This tradition's institutional networks are based around five temple-centers (Hindi: *math*), one of them being Kedarnath, and five world teachers (Hindi: *jagatguru*), of whom the Kedarnath *jagatguru* (who also bears the title *rawal*) Bheemashankar Ling, who lives in Ukhimath, is one. His ordained students (Hindi: *cela*) serve as *pujaris* in Kedarnath.

THE ORIGINS OF THE GREATNESS OF KEDARNATH

For at least several centuries and potentially much longer, Kedarnath has been a place connected to a web of stories and ritual practices that involve Shiva's presence and absence, the human pursuit of God and self-perfection, the transition from life to death, and the imperfections built into being human and trying to act in ways that are just, righteous, and true to your own nature—that is to say, according to *dharma*. These stories and ritual practices are set in a location that bridges the world of humans to worlds beyond the human. Any specific telling of one version of a Kedarnath-related story eventually resonates with other parts of this web. Several intertwined and contextualized anecdotes will show what I mean.

On June 29, 2007, in Kedarnath, Bhupendra brought a Kedarnath pilgrimage priest whom I will call Tiwari-Ji to our shared room. Unlike many people with whom I had had conversations only because of my own or Bhupendra's persistent effort, Tiwari-Ji sought us out. He found Bhupendra in the bazaar and told him that he would like to meet with me to make sure that I had the correct understanding of the origins of Kedarnath. Bhupendra brought him to our room and we recorded the conversation. The following excerpted telling is worth attending to in detail because it frames many of the key issues for necessary for understanding how people view Kedarnath both inside and outside the region of Garhwal. Tiwari-ji began by referring (initially without explanation) to one of the most famous details of this story: that Shiva tried to hide from the Pandavas by taking the form of a buffalo.

> There are different published versions coming out—there is a Hindi one, there is a Marathi one, here there is a Gujarati one—it depends. . . . I will tell you particularly about the subject of Kedarnath-Ji, what is the importance of Kedarnath here, what is the importance of God here. So many people come here—why do they come? With what sort of mind-set are they coming? Some people come here and say that the true form [Hindi: *svarup*] of God [Hindi: *Bhagwan*] is that of a buffalo. A buffalo, meaning, they say that . . . the buffalo is God's true form.[8] But, and here I give something of my own presentation that I give to *yatris,* to them I say that the true form of God is not in the form of a buffalo. God's *linga* of light [Hindi: *jyotirling*] is here from before then. That is to say, God's *linga* of light is here from before the beginning.

Here Tiwari-Ji mentioned two other stories connected to Kedarnath that he regarded as notable: the time when Ravan took *darshan* of Shiva in Kedarnath, and the time when Shiva told a king named Kedara that, as a boon, he would add the suffix -*natha* (Sanskrit: lord, in Hindi -*nath*) to the king's name and make it into the place-name for the Himalayan abode we now know as Kedarnath. Tiwari-Ji then went on:

> We can say that there is no single definite story that we can tell you here that says, "Brother, this is God's importance, and this is the greatness of the place." Starting a long time ago, God has given *darshan* [or alternatively "become manifest"] in different forms [Hindi: *rup*] and in different ways [Hindi: *dhang*]. You will have already obtained some information on the history of the Pandavas; the history associated with them is from almost five thousand years ago. Bhagwan himself knew that this *linga* of light is without beginning.

As noted above, part of the traditional responsibility of pilgrimage priests like Tiwari-Ji is to offer a coherent understanding of the character of the pilgrimage site, usually in the form of a story. For Kedarnath, as for almost any other place of religious significance in the world, there are always multiple stories, and what the *tirth purohit* tells on any given occasion is typically a context-sensitive and synthetic performance. In the beginning of his narration, Tiwari-Ji made it clear that he was aware of possible contradictions among different foundation narratives about Kedarnath, possible confusions about Shiva's form in the place and how that form came to be established. He began by harmonizing the idea that the greatness of Kedarnath stemmed from the arrival of the Pandavas with the famous account of Kedarnath as a *jyotirlinga,* a *linga* of light whose presence in the place is without beginning, the aspect of Kedarnath that is the subject of the next chapter, "Lord of Kedar." As Tiwari-Ji related,

> Lord Shri Krishna knew that there is a *jyotirlinga* here and that until the Pandavas took *darshan* of that *linga* their wickedness [Hindi: *pap*] would not be cleansed. . . . After the *Mahabharata* war happened, a very awful war that lasted for eighteen days in which everyone who was related to the lineage of the Kauravas died with no survivors, including the Pandavas' guru . . . their blood relations, those who were special to them—there were no survivors—at that time, when at the end the Pandavas finished ruling their own kingdom, they felt like this: "What did we do, and who made us do this? Lord Krishna made us do this. So much grievous calamity!"
>
> So Yudhishthira said to God [here, Krishna], "Lord, you have caused us to be so very guilty [Hindi: *hamara itna bara dosh lagaya*], guilty of killing fathers, guilty of killing mothers, guilty of killing those of our lineage, tell us what greater *pap* [Hindi: wicked deed] in the world is there? Where do we need to go for expiation [Hindi: *nivaran*]?" So, Lord Shri Krishna said, "Actually you have done really wicked deeds [Hindi: *aghor pap*]." When Lord Shri Krishna said this, then Dharmaraja Yudhishthira himself said, "It was you who inspired us, saying, 'Fight, fight, destroy *adharma*' [Hindi: anti-*dharma,* in this context injustice]. And now you are saying to us, 'You have done wicked things.'" So then Lord Shri Krishna said, "It was against *adharma* that I inspired you, but just because it was against *adharma* does not mean that you were not fighting. How would that happen? *Adharma* always grows, it never lessens, it never ends. But you actually did the deed—you killed relatives and you killed your guru, so you are guilty of murder. And for that you need to perform expiation."
>
> So Lord Shri Krishna again said to them, "Go to the Himalaya, go where God [Hindi: *Bhagwan*] has a place called Kedar. God has a *jyotirlinga* there and he himself is present [Hindi: *virajman*] in that place, and until that *linga* appears to you there and until you touch [Hindi: *sparsh*] it then you will not obtain the actual [Hindi: *sakshat*] *darshan* of God." So the Pandavas wandered and wandered until they came to Guptkashi [in the Kedarnath valley]. Lord Shankar [Shiva] and Mother Parvati were resting there, and Mother Parvati was troubled that the Pandavas were distressed and worrying, "Where has he gone, has he gone in this valley or that valley. . . . Where is this road?"

The body of Tiwari's account begins near the end of the plot of the *Mahabharata* epic, after the war in Kurukshetra has finished. The Pandavas are victorious, having followed the counsel of Krishna to return from exile and recover their kingdom from their relatives through a war so bloody that James Hegarty (2012, 78) termed it "megadeath." Yet what they did in the pursuit of *dharma* was morally complicated and painful, and it weighs heavily upon them. They killed friends and kin. In Sanskrit versions of the *Mahabharata* a heavy grief hangs over the story.

As his male subjects "slaughter each other in an orgy of alliterative, drink-fueled, violence," the god-in-human-form Krishna dies (Hegarty 2012, 77). Only after his death do the Pandavas begin a *yatra* up into the Himalaya for what Hegarty (2012, 77) terms "suicide-by-pilgrimage." Tiwari-Ji continued:

> Whenever one goes to a new place . . . it is necessary to ask people where is this side street, where is this store—that's how it is in cities. So at that time what must their experience have been like? . . . It must have been dark, *no population* [said in English], no habitations [Hindi: *basti*], just jungle. You just take a walk around here five thousand years ago. . . . All these hotels and lodges you see were built since I've been living here, in the last ten-fifteen years. Before that there were . . . huts here. . . . People used to make huts or temporary dwellings [Hindi: *jhompri*].

This story, told in 2007, echoes many conversations I had in Kedarnath with both locals and visitors where the speaker would compare the high-intensity, overbuilt Kedarnath of recent years with that of earlier and quieter times when the place was in more accord with its Himalayan surroundings and when the signs of human presence made no pretense about their ephemeral nature. The year 2007 in Kedarnath was, like much of the last decade and a half, a time of constant building. People were renovating old structures and adding on new rooms and new floors targeted at the increasingly middle-class *yatris* who were coming to Kedarnath willing to pay for larger rooms with in-room hot water, fancy blankets, and prepared-to-order regional cuisine. There were always contractors bringing in cement on the backs of ponies. It was a time when it made sense to invest, and increasingly Kedarnath had become a testament to the incentives of capital—a bustling small village at approximately 3,500 meters (almost 12,000 feet) in the Himalaya where different kinds of food, clothing, movies, medical services, forms of transportation, cell phone service, a railway reservation office, and related amenities were usually available. It added to the impressiveness of the place—that humans had been able to create so much so fast in such a challenging environment. The farther back one goes in the history of Kedarnath in the twentieth century, the less common it was to actually spend the night in Kedarnath. Tiwari-Ji continued:

> In Gaurikund, here in Kedarnath, *yatris* used to live in those huts, so if you took a walk around even before that what would be there? In my opinion, there would have been . . . jungle, forest, well there's not that much jungle in the Himalaya, so there would have been snow. So, Lord Shankar [Shiva] says to Mother Parvati . . . their goal is there in the Himalaya, at Kedarnath, they have to search for the *linga* at Kedarnath [i.e., not in Guptkashi]. So Mother Parvati says, "You do what you want, but I will certainly give them *darshan*." So Lord Shankar disappeared—there in a manner of speaking he became hidden [Hindi: *gupt*]. And they say that this is where Guptkashi [Hindi: hidden Kashi] gets its name, and there [in Guptkashi] Mother Parvati herself gives *darshan* to the Pandavas. And she says to them, "I can't do

anything for you now, but I can give you a little guidance and tell you where to go, which road to take—I can do that much, I can make the road comfortable for you." So when Mother Parvati said this they somehow saw the road here, and then they saw that buffalo, that illusion in the shape of a buffalo [Hindi: *maya-rupi bhains*], they saw that here.

With this mention of the buffalo, Tiwari-Ji began to narrate in detail one of the distinctive stories about Shiva's presence and form in Kedarnath: that Shiva had taken the form of a buffalo to hide from the Pandavas. In Tiwari-Ji's view this decision was itself a memory of an earlier occurrence: "Once it happened that . . . God took the form of a buffalo to kill the demon Bhasmasur here. . . . It was that buffalo form that God assumed here." Then Tiwari-Ji returned to the historical present of the story in which the Pandavas were trying to find where Shiva was hiding among the buffalo herd.

> So . . . he [probably Yudhishthira] said to Bhima, "Make those buffaloes come out [Hindi: *nikalo*] from the herd one by one; among them is the true form of God." So as they were sorting the buffalo herd one by one, each of the buffalo went to the *linga* [which as Tiwari-Ji previously mentioned was already present in the place] and were getting absorbed in it [Hindi: *us hi ling mem ja rahe hain samavesh ho ja rahe hain*]. The Pandavas . . . are driving the buffaloes the way they drive cows [Hindi: *hamkna*], one by one. . . . So the buffalo that was God, it came here and right at that moment it got absorbed into the Shiva *linga*.

Shiva hides from the Pandavas. He takes the form of a buffalo, a form more commonly associated with *rakshasas*, demons whose battles with gods and goddesses constitute some of the most enduring and foundational stories about the gods and goddesses worshipped across South Asia. From this buffalo form the Pandavas drive him into his more fundamental, beginningless form—the *linga* that is already part of the landscape of the place. In this narrative, Shiva is both already in the place and coming to be present in a new way. In many accounts of this scene Bhima spreads his legs and forces the buffalo to pass through his legs one by one— a passage that Shiva refuses to make, thereby identifying himself. Work by David Shulman (1976) has shown that in South Indian versions of conflicts between Devi and the buffalo demon Mahishasura the buffalo demon is sometimes identified as a devotee of Shiva; sometimes the vanquishing of the buffalo demon also entails direct violence by the Goddess toward Shiva himself. Stephen Alter (2001, 265) has suggested that this moment in the Kedarnath story is connected to a moment in the rituals surrounding the historically common practice of buffalo sacrifice (usually to a form of the Goddess) in Garhwal and Kumaon in which it is made certain that none of the buffaloes to be sacrificed are gods in disguise. This confirmation is carried out by forcing the buffaloes to pass through the center of a broken rock. Alter suggests that in the Kedarnath story Bhima's legs take the place of the broken rock. Thus this scene in the Kedarnath foundation story may signal Bhima's intention to carry out a buffalo sacrifice, a threat that mythically conflates Shiva's buffalo form with the buffalo form of his demonic adversary.

Tiwari-Ji continued: "So one by one they were disappearing, every single one of those illusory buffaloes disappeared, and the one that was God's true form began to be absorbed into that *linga,* it was going there. Like when you have seen in some documentary. . . . That all happened. . . . Bhagwan's buffalo form was being absorbed into the *linga,* it was showing the way."

In many versions of this story, the Pandavas at this point identify Shiva and grab him to prevent him from leaving. Each of the five brothers grabs a part of Shiva, parts that remain in the landscape and then become the self-manifest rock *lingas* found in Kedarnath and the other four of the Panch Kedar (the Five Kedars, five Shaiva temples in central Garhwal). Shiva's backside, his "back-portion" (Hindi: *prishth-bhag*), becomes the Kedarnath *linga.* In other versions, Shiva's *prishth-bhag* stays in Kedarnath and his face emerges at Pashupatinath in Kathmandu, Nepal.

> So when the Pandavas grabbed [Hindi: *pakarte hain*] that buffalo and touched it, in a manner of speaking they touched the *linga* of God [here Shiva]. They achieved . . . the touching of God; the true form of God that is present in the light is what the Pandavas touched. Then you may understand that the buffalo disappeared and then there . . . it happened that the divine form got extended/stretched [Hindi: *jo divyasvarup tha vo tan hua*] and God in original true form gave *darshan* to the Pandavas. And he said, "Look, you endured pain and you have also done wicked deeds, but however much wickedness you did now it has all been cleansed [Hindi: *dhul ho gaya hai*]. Now tell me what you wish for—what do you desire?"

Once, during an interview I held in 2005 with a family living in Jaipur's old city who had visited once Kedarnath, an elderly woman strongly disagreed with the idea that the Pandavas had "grabbed" Shiva. The verb *grab,* when used in connection with cattle, can refer to the common way of steering cattle by grabbing them near the top of the tail (Hindi: *punch pakarna*) and twisting the tail to make the animal move.[9] It was a complicated moment—I had a printed pamphlet version of the story with me at the time that laid rest to the worry that the discrepancy had been solely due to my own insufficient level of Hindi knowledge. But when I showed her the printed version she, and the rest of her family, became uneasy. It felt difficult to them. How could the Pandavas have been disrespectful to God? They would never have done something improper (Hindi: *anucit*).

Because of this conversation, the question of the propriety of the Pandavas' actions became part of my stock set of conversation and interview questions when speaking to people about Kedarnath. In some versions of this story Bhima, the famously too-passionate and fierce Pandava, is angered by Shiva's unwillingness to be fully present for the Pandavas and strikes him with his mace, a narrative variant that underscores the ambiguity of whether and how we are to view the Pandavas at this moment in the story as moral exemplars. On May 3, 2007, when I was sitting in the reception area of the Maharashtra Mandal *dharamshala* in Kedarnath with several families from Maharashtra and Rajasthan, I asked one of the husbands from this group whether he thought Bhima's behavior was appropriate. The question gave him pause. His wife said she thought Bhima's behavior was acceptable because he was trying to obtain liberation (Hindi: *mukti*) and therefore the means were justified. The implication, made explicit in many other such conversations, was that such behavior toward a deity would normally not be appropriate. When I was standing in line on May 20, 2007, and discussing this story, a member of a group from Gujarat said that the Pandavas were after all only human and that humans get angry (Hindi: *gussa*), even if they are standing in front of God.

The quest of the Pandavas for Shiva does not feel like a straightforward quest. It is the story of a deity who makes himself intentionally difficult to reach by humans whose search for him shows just how human they are. Now we return to Tiwari-Ji's account:

> Then the Pandavas said, "For what purpose would we need a favor? We just need a little place to live in the Himalaya—-we just want to live here." Then the Pandavas felt this inspiration and craziness inside that "we should build a temple and do God's *puja*," and God said, "Do your *puja*. And . . . make a road for coming generations, since you have already made a temple. And when future generations come, among them your names will be immortal. Whenever any *yatris* come here in God's name they also will come in your name." And the Pandavas' name became immortal.

Kedarnath, as Shukla-Ji noted in the Introduction, is a meeting point for mountain people, particularly of the Kedarnath valley, and *yatris* from all over India and beyond. It is possible to see the echoes of this idea in how Tiwari-Ji speaks of the Pandavas here. His presentation combines the internal regional importance of the Pandavas with their broader reputation beyond the region that is known to visitors. As William Sax (2002, 43–44) and Karin Polit (2008) have shown for Garhwal, and as Jon Leavitt (1988) has shown for Kumaon, many residents of Uttarakhand understand their mountains to be the birthplace of many of the primary characters of the *Mahabharata,* both Pandava and Kaurava, and locate much of the plot of the epic in their own region. For locals, particularly belonging to Rajput castes, the Pandavas and in some cases Kauravas such as Duryodhana are ancestors, effectively lineage deities whose stories are performed every year and whose powerful personalities possess the performers. Looking at the large stone blocks out of which the current temple is built, a structure that later withstood the floods of 2013, one could easily imagine that the Pandavas who managed to bring these massive pieces of stone up into the mountains were more than human and worthy of worship as lineage deities. The geographer Surinder Bhardwaj (1973, 146) famously created a typology of Hindu pilgrimages based on survey data about the relative distances between home place and pilgrimage place traveled by *yatris*: "local" sites patronized only by residents of a specific locality, "subregional" (high and low) sites important to a related set of local communities, "regional" sites visited by devotees from across a region, "supraregional" sites that pulled visitors from across regions, and "Pan-Hindu" sites whose attractive power rose beyond even the supraregional. From this typology it becomes clear that the connection of the Pandavas to Kedarnath is one of the factors that makes Kedarnath significant at almost all of these levels.

At this point Tiwari-Ji's account turned to the Great Path:

> Then the Pandavas went ahead, and perhaps they disappeared somewhere in the Himalaya. They say about Dharmaraja Yudhishthira that he was journeying in the Himalaya, and one time there was a dog going with him, *dog* [said in English], *shvan* [the Sanskrit word for dog as opposed to the Hindi *kutta*]. The dog was going with him. It was a big surprise to Yudhishthira: "There is a dog going with me. Where did this dog come from? There aren't any habitations close by, so why he is going with me?" The dog was going where he went. So at one point during that time, they say, the dog got worms. It would be a surprise, how do you get worms in the Himalaya anyway, but they say that Dharmaraja Yudhishthira took those worms out of the dog and threw them on the ground. . . . [Then] this also seemed like a wicked deed [Hindi: *pap*] to him. Yudhishthira thought, "I already committed *pap* on my way here. Those worms shouldn't die and also the dog should survive." So they say that Dharmaraja Yudhishthira tore open his thigh and taking the worms out he put them into his own thigh. Then God himself gave him true *darshan* and said, "I thought that they just call you Dharma-King [Sanskrit: *Dharmaraja*], but actually you really are a king of Dharma." And it is believed that Yudhishthira went to heaven in his own body.

In between the first and second sentences of this part of Tiwari-Ji's exposition lies the particularly poignant segment of the journey of the Pandavas into the Himalaya that I saw on a television screen while I waited in line to enter the temple. When they enter the Himalaya the Pandavas number six: Yudhishthira, Arjuna, Bhima, Nakula, Sahadeva, and their joint-wife Draupadi. But as they walk further up in the Himalaya, toward the heavens of the gods, one by one everyone but Yudhishthira dies in some way, and their deaths are clearly linked in many tellings to a moral failing from their life. By the end, only Yudhishthira, the character of the *Mahabharata* who is defined by his pursuit of *dharma* regardless of the human cost, remains. His guilt at the death of the worms in this telling exemplifies this trait. Yudhishthira does not die: he walks into the afterlife *in his own body* (Sanskrit: *sadeha*). This is the path of what Jvalaprasad Mishra, in the foreword to one of two commercially published editions of the *Kedarakalpa* (a late tantric text connected to Kedarnath) calls "the path of going to Kailasa in one's own body" (Hindi: *sadeh Kailash jane ka marg*) (Padumā and Hajārībāg 1907, 1). This idea of liberation through the body, often involving a process of self-divinization, is one of the hallmarks of the tradition of South Asian practice that is described by the umbrella term of *Tantra*. The *Kedarakalpa* tells the story of the journey of five yogis who travel from the world of death into the presence of Shiva at Mount Kailasa and into the state of liberation (Hindi: *moksh*) via the Garhwal Himalaya. They reach this state of liberation, also glossed as the state of being in the presence of Shiva, with "their physical bodies intact," an important difference from how nontantric experiences of death and liberation are often conceived (Goswamy 2013, 190). As the *Kedarakalpa* narrates this journey it also describes the sacred potency of the Himalayan landscape in minute detail and provides instructions for different ritual actions (drinking water, reciting powerful sacred verses [*mantras*]) that are to be done at different locations.[10] I will return to some of these details in the next chapter.

The area in and around Kedarnath is associated with death and what is beyond death in several related ways. Several scholars regard it as quite likely that the site now known as Kedarnath may have been once been better known as "the high place of the sage Bhrigu" (Sanskrit: Bhrigutunga) or "the falling place of the sage Bhrigu" (Sanskrit: Bhrigupatana), which is mentioned in the *Mahabharata* (Naithānī 2006, 175–76; Dabarāl, n.d., 61–62; Rāvat 2006, 84). Garhwali historian Shivaprasad Dabaral suggests that the "Path of Bhrigu" (Hindi: Bhrigupanth) is another name for the Great Path and notes that the *Mahabharata* records that Arjuna went there on his journey through the Himalaya (Dabarāl, n.d., 308–9). William Sax (1991, 28), summarizing the work of historians and colonial-era travelogues, writes that "the Path of Bhrigu above Kedarnath . . . was traditionally associated with suicide." This life-ending practice would have taken the form of devotees jumping or falling from a precipice in or near Bhrigutunga/Bhrigupatana, and Dabarāl (n.d., 308) connects this practice with the early identity of the site of Bhrigutunga (Dabarāl, n.d., 308). Sax (1991, 22) recorded a song sung by Garhwali women from the village of Nauti to the goddess Nanda Devi in which one of her forms, Maya, ends her life by jumping from "the high path of Bhrigu" because she is distraught over her inability to find a husband who can complete her and fulfill her desire.

I remember in 2007, immediately prior to the opening of Kedarnath for the pilgrimage season in May, there were rumors in the Ukhīmath bazaar that an elderly man had managed to make his way all the way to Kedarnath village without anyone noticing, had walked up into the end of the valley immediately behind the temple, had sat down there, and later was found dead in that position by police. When one heads into the high Himalaya the distinction between physical ascent and spiritual (Hindi: *adhyatmik*) ascent seems to blur.

THE LINE

All of the above would be in my mind when I waited to enter the Kedarnath temple and take *darshan*. I would reflect about what it meant to say that in Kedarnath one stands in the footsteps of the Pandavas. Their pursuit of Shiva is a morally ambiguous story—in some versions they win access to God through an irreverent persistence that verges on violence, while in others they reverently come before Shiva at the end of a long process of self-purification. Shiva is everywhere in the story: in the ground, as the buffalo, and as universally pervasive vibratory light. Always we know that afterwards the Pandavas walk up and out of the human world. During the high season in 2007, 2008, and 2011, the dynamics of the line reminded me of the moral questions raised by the Pandavas' ambiguously reverent and irreverent pursuit of Shiva. I came to see the line, or queue, in Kedarnath as the site of an embodied controversy about the relationship of means and ends. In South Asian contexts, analyzing the queue, and more broadly the act of waiting, as a situation filled with symbolic weight and social contestation is of course not a novel idea (Jeffrey 2010; Gandhi 2013; Corbridge 2004).[11] But in Kedarnath, as is common in large pilgrimage places in South Asia, waiting in line was one of the events that defined the public life of the place much of the time, and the story of the Pandavas explicitly frames the place, at some level, as a place for thinking about *dharma*. So as I stood in line myself in Kedarnath, I came to see how this well-known situation could be seen to take on something of the character of the place. Some *yatris* would attempt, either surreptitiously or brazenly, to jump the queue and enter the line wherever they could. For them, what mattered was access to what and who was inside, however gained. After all, just as in the case of the Pandavas, in the pursuit of *mukti* what is not justified? This would enrage others who felt that putting up with the hardship of waiting in line was part of the necessary internal purification for going in front of Shiva in a state that was pure (Hindi: *shudh*). Verbal and physical confrontations were commonplace during the line in high season. These confrontations also implicated local tensions about control of the site and access to the temple. I do not mean to suggest that this framing was shared by most of the people who waited in line with me. But I think it helps us to see how tightly embodied experience, story, ritual practice, and meaning are bound up together in such moments.

THE RESEARCHER WAITS IN LINE

I often made a point of waiting in line myself rather than simply observing the line. Many Kedarnath locals did not understand why I did this. It was clear that, both because I had been already living in the Kedarnath valley for about five months by the beginning of the 2007 and because I was a white male American researcher,

I could enter by the side door whenever I chose. However, I usually waited in line. As a participant observer I wanted to experience the line, and it was usually a useful opportunity to chat with *yatris*. I also did not want to draw the ire of *yatris*, and personally I have never liked people who jump ahead in lines. But waiting in line sometimes turned out to be a trying experience. It was often a moment when circumstances would force me to wrestle with the complicated and challenging issues bound up with being who I was and working in the Kedarnath valley. The following is an example of what I mean that ties together some of the themes and information I have introduced in this chapter.

On July 3, 2007, in Kedarnath, I had fixed an appointment to interview the chief (Hindi: *pradhan*) of a village in the Kedarnath valley. It turned out that I had, unknowingly, fixed an appointment to be interviewed. In a dark room in a lodge just off the main bazaar, in front of several other local Kedarnath valley men, my interviewer disingenuously begged me to write about the glorious Kedarnath that is described in Puranic literature and not to write about the current state of the place. Another man wanted to make sure that I was not doing some kind of spying (Hindi: *jahsousi*) that would later result in a sociological exposé. This reverse interview was a nerve-wracking conversation that went on for over half an hour on this single point. I refused to simply promise to do as requested. I said that if I promised but then went on to describe what was really happening in Kedarnath the chief would say that I had broken my promise. I said that my job, as a researcher, was to write about the relationship between what people experienced (Hindi: *anubhav*) in Kedarnath and the realities (Hindi: *hakikat*) of what the place felt like from day to day. I said that this meant that I would in fact describe, to some extent, some of the messy social realities that could be found in Kedarnath. But I continued to say that, as a scholar trying to describe *hakikat*, I would be doing a poor job if I did not also attend to the power of Shiva understood to center in the temple. Everything in Kedarnath directly or indirectly connected to the presence of Shiva or the Himalaya or the Goddess or some combination of these. At this point my primary interviewer asked me to describe my own experience (Hindu: *anubhav*). So I told him a story that began with my waiting in line to enter the temple.

On May 30, 2007, my father was scheduled to have a cataract operation, so I decided that I would have a *puja* done the day before. I was feeling my distance from my own family at that time very keenly and thinking that it could not hurt. I joined the line at 5:30 a.m., a half hour before the temple doors opened for general *darshan* at 6:00. When I neared the door about two and a half hours later I was extremely annoyed. My patience was gone, my ostensible distance from the situation as a participant observer had vanished, and my temper was frayed to the breaking point by the numerous times I had felt compelled to verbally and sometimes physically prevent people from entering the line either directly in front of me or directly behind me. The worst was when a woman approached me and announced that she was rejoining the line after putting down her sandals outside the temple courtyard. She entered the line directly behind me and steadfastly stuck to her lie in the face of all my protests, then succeeded in getting into line five spots behind me. The second worst was when I was offered money, which I angrily refused, to let people enter the line directly in front of me.

As I entered the door to the temple I felt very troubled: In such a state, what is the use of doing a *puja,* even as a non-Hindu participant observer? This doubt grew as I proceeded in line through the anteroom and up to the doorway of the inner sanctum. Then everything changed. At the same moment, when I saw both the Kedarnath *linga* and the *Ved-pathi* who would do my *puja,* my annoyance and anger somehow evaporated, and for reasons I still cannot parse I began, against my will, to weep. Some of what I was feeling at that point, as I listened to the mantras said by the *Ved-pathi* and clutched his knee as I fought for position in the crowded space, was a very strong memory and love and concern for my father, and the rest of what I was feeling I simply cannot name. I then left as quickly as I could, did circumambulation (Hindi: *parikram*) of the temple, and went back to my room without speaking to anyone and stayed there for several hours, in the dark. This story changed the conversation. The immediate response of my interviewer was that I had experienced *sakshat darshan,* the vision of an ultimately true form of divinity. This was a bit of ethnographic irony: it had become standard for me to ask about *sakshat darshan* during conversations and interviews (whether someone believed it was possible, whether the particular person had experienced it). Here it was my own connection to *sakshat darshan* that was the topic of conversation. I said that I had not felt, during my experience in the inner sanctum, that I was seeing Shiva in a true way. I simply had had an emotional response that I did not understand and could not control. He said, "You do not realize that you had *sakshat darshan,* but that is what it was." Shortly afterward, the conversation ended.

MY PLACE AS A RESEARCHER

Not only this conversation but many others with pilgrims and locals raised questions for me about where to place myself, and where others placed me, on the already existent continuum of ways that Westerners were present in the area. Anthropologists had worked in the Kedarnath valley before (notably Karin Polit and Eric Schwabach), and Kedarnath valley residents had often heard of William Sax. Westerners with highly variable levels of cultural sensitivity and cultural literacy came through the Kedarnath valley area as tourists and trekkers all the time and stayed anywhere from several days to several months. A German woman, formerly a student of anthropology, was drawn to the Kedarnath valley through her dreams. She became the disciple of a renunciant guru in the Kalimath area, learned Hindi, Nepali, Sanskrit, and Garhwali, and achieved a high level of respect in the area. Now referred to as a mother (Hindi: *ma* or *mai*), Saraswati-Mai, as she is called, has been in the area for well over a decade and carries out *puja* at her guru's ashram. There is an international nongovernmental organization (NGO) based in Ukhimath. Employment with this NGO is highly prized, and it was with considerable consternation that I found myself renting two rooms in a house that had been historically connected to it. I did not want to be seen as another rich American deepening the groove of Western economic involvement with the Kedarnath valley region. Yet I was unable to avoid this trajectory completely, and Bhupendra worked for that very NGO after working with me. The Kedarnath valley also saw the occasional Western renunciant.

Usually I presented myself primarily as a researcher—a somewhat familiar but also murky category. Introductory conversations would often follow a pattern: Are you a tourist/trekker? No. Are you with the NGO? No. Ah, then you must be doing "research and stuff" (Hindi: *research vagaira kuch*). Yes, I would say. I once asked a teenage boy what he thought *research* meant. He said that it meant searching for things (Hindi: *khoj karna*), which I thought was a reasonable answer. My goal was in many ways ironically parallel to that of the *tirth purohits*: I wanted to create rapport with visitors during their short time in Kedarnath so that I could collect something from them. My primary positionality with regard to *yatris* was not that of scholars such as Ann Grodzins Gold (1988) or E. Valentine Daniel (1984), who as researcher-participant went on a *yatra* for the duration of the journey with a group of *yatris*.

Yet in other ways I was myself a *yatri*—a visitor to the region struggling to balance my reactions to the awe and power of the place with the mundane difficulties of living at an altitude of approximately 3,500 meters without having spent most of my life in a similar situation. As a participant observer trying to be as much a part of what was going on around me as possible, I looked like an odd sort of Shiva devotee (Hindi: *Shiv-bhakt*). At one point during my time in Kedarnath itself, an important officer of the Samiti at Kedarnath asked me, in the presence of several others, when we were going to go ahead and do my sacred thread (Sanskrit: *upanayana*) ceremony to become Hindu. I declined, saying that if I did the *upanayana* ceremony it would complicate my life in America and my family would not like it. When people would ask me why I was doing what I was doing, I think my answer that satisfied all parties the most was that I was studying how they experienced the place and trying to understand my own attraction (Hindi: *akarshan*, recalling Shukla-Ji's use of this idea in the Introduction) to it at the same time.

I experienced a kind of *khicri* (Hindi: a dish where rice and lentils are cooked together along with whatever else one has at hand) intersubjectivity to working in Kedarnath. My interactions with visitors to the region were brief encounters whose tenor was widely variable. Sometimes they would be quick, surface conversations, but on occasion a deep rapport arose by virtue of the conditions we shared: on *yatra* in a special place, far from home, high up in the mountains in fickle weather conditions. Yet I did, in my own complicated foreign way, partially join the fraternity of those who worked in Kedarnath. *Fraternity* is a fair word, I think. Kedarnath closes down entirely for the winter, and during the summer months of the pilgrimage season 99 percent of the temporary residents of Kedarnath are men. Kedarnath is not a socially three-dimensional Garhwali village. There are few women and children. There are no schools or homes with family *puja* shrines. The pipes and phone cables are put up and taken down each season.

In contrast to my relationships with visitors, I developed long-term relationships and friendships with some Kedarnath valley locals and, conservatively, am known to hundreds of people connected to Kedarnath in the Kedarnath valley. But there has always been a complexity to developing these relationships—the differences of history, culture, and socioeconomic status hang between us in ways that have been very difficult to move past, I think in part because of the intensity of the processes of commodification that have surrounded the pilgrimage tourism industry in recent years in Garhwal. Thus my work with Kedarnath has not been based on long-term relationships cultivated with individual women in the manner of Karin Polit (2012) or the decades-long collaboration of William Sax with Dabar Singh or of Ann Grodzins Gold with Bhoju Ram Gujar (Sax 2009, vii, 7; Gold and Gujar 2002, 30). Bhupendra and I had begun to move past the foreigner-guide friendship where we began, but he is not here anymore. In and around Ukhimath, as I became more familiar to locals and became myself more familiar with how to communicate and navigate, I began to have substantive conversations with Garhwali women, but this really began only after I had been working in the Kedarnath valley for well over a year. Once or twice near the end of my dissertation fieldwork in 2008, as I walked down a lane in Ukhimath toward Omkareshwar Temple (where Kedarnath-Shiva is worshipped in the winter), several children called me "uncle" in Garhwali, a sign that I had entered their social map. But woven through all of these long-term and short-term relationships was my persistent focus on and engagement with *place*: Kedarnath, of course, but also Ukhimath, Guptkashi, Lamgaundi, and Madmaheshwar as well as several other villages and shrines in the Kedarnath valley.[12]

The complicated nature of these local and visitor relationships was in a way a sign of the times. At a pace that correlated roughly to the growth of pilgrimage tourism in recent decades and the new statehood of Uttarakhand (first as Uttaranchal) in 2000, the Kedarnath valley had begun to leave behind the Garhwal chronicled by William Sax, a Garhwal in which locals were conversant with their performance traditions and unashamed about the powerful, possessive presence of their local deities. The performance of episodes from the *Mahabharata,* possession by local forms of the deity Bhairavnath, all-night singing and drumming rituals that worshipped local deities (Garhwali: *jagar*), and other features of Garhwali culture were still found in the area, but many young people were beginning to have complicated and often uncomfortable re-lationships to these cultural forms. Some found possession a bit embarrassing, and most could not understand or repeat the words to songs sung in classical/older (Garhwali: *tet*) Garhwali. As Karin Polit (2008) and Stefan Fiol (2010, 2012) have documented, these older performance traditions have begun to pass into different performative and representative registers: as stand-alone pieces in heritage festivals, staged performances filmed for commercial video distribution beyond the region, or private videos uploaded to YouTube and shared on Facebook. There is a genre of Garhwali songs known as "high-pasture songs" (Garhwali: *payari geet*) that village women sing to welcome traveling deities who, carried on palanquins, have come down from the high places where they live and are traveling through their territory and visiting the villages there.[13] In the Kedarnath valley women, reportedly, used to sing these songs for the Kedarnath deity procession but do not do so anymore. For Madmaheshvar, another locally important form of Shiva, there are still women who sing *payari geet* to welcome him, but few women know the words today and most are middle-aged or older. During my fieldwork I have been asked to explain the plot of *Anaconda* as it was playing on HBO in the Ukhimath bazaar and the refrain to Shakira's "Hips Don't Lie." I once entered the living room of a friend's house to find his grandmother watching *World Wide Wrestling*. Standing in line and seeing a video showing the Pandavas on the Great Path reminded me of these traditional Garhwali religious worlds. But the fact that it was a video, and that there were screens in Kedarnath capable of showing such videos (at least when the electricity was working), also reminded me how much was changing. And my own fraught experience of waiting in line reminded me of my own complicated relationship to the people around me and to Kedarnath.

FRAMES OF APPROACH

The story of the Pandava's climb to heaven makes a good narrative frame for approaching Kedarnath generally in the context of this book. This suggestion builds on a mode of interpreting myth suggested by Laurie Patton and Wendy Doniger (1996, 392): that scholars should "take myths seriously in their own right . . . evaluate them as forms as narrative reasoning equal to our own." While this approach is, unmistakably, my own analytic imposition on the data I have gathered, I think that it is a reasonable analytic imposition that arises out of some of the most important and best-known stories about Kedarnath. We see in this tale a sense of Kedarnath, and more broadly its Himalayan locale, as a place of human effort, of pursuit and violence. The journey to Kedarnath is not a journey that everyone survives. We see the ways in which Kedarnath is a place that both demands preparatory self-cultivation and puts people in positions that reveal their human failings. It is a place that both cares and does not care about human behavior. It is also a place, as I shall discuss in the next chapter in more detail, that enshrines Shiva's simultaneous presence and absence.

From conversations I have had with locals in Kedarnath about the idea of occupation (Hindi: *dhandha*) as it relates to the central and difficult-to-define idea of *dharma,* which in this context I will translate as the virtuous way one ought to act based on one's identity and/or occupation, it is possible to see the ambiguity of the connection between the power of Kedarnath and the necessity for proper behavior. I often had conversations in which people told me that their *dhandha* required them to have sharper dealings with guests and patrons than they personally would prefer, a trend that had clearly intensified in recent years. The amount of money that was circulating in the Kedarnath of recent, predisaster years was changing how people made decisions and how they related to one another. Now this has become a matter of regret, felt by some to have contributed to the disaster. Yet what does it mean to behave improperly in a *tirtha*? On the one hand, it is particularly inappropriate. On the other, *tirthas* are places that, by definition, have a surfeit of purificatory power, the ability to cleanse even the worst *karma*. We see both sides of this in the story of the Pandavas' climb to heaven (Hindi: *swargarohan*).

The interpretive depths of this story, however, should not create the impression that I think Kedarnath should be counted among what Elliott McCarter (2013, 50), in the course of his work on Kurukshetra, has termed "narrative-centric sites and sites where narrative formulations are active and meaningful." The stories of the Pandavas frame Kedarnath, as does the popular conception that the Himalaya are the preferred abode of Shiva. Yet it would be a mistake to understand Kedarnath as a place where the narration of these stories stands at the center of the character of the *tirtha*. Indeed, in recent years before the floods, I noticed that during the high season many *yatris* would come and go from Kedarnath without having heard a full version of this story, a development that, as Andrea Pinkney (2013a) has observed, makes the printed pamphlet versions of these stories in English and Indian regional languages all the more important. There were simply too many people and not enough time for what in earlier times would have been the ritual of narration by the *tirth purohit* soon after arrival into Kedarnath. Kedarnath is also not a place that has grown up around the idea of the investigation into and practice of virtuous conduct, as Leela Prasad (2006) has shown for the town of Shringeri, which is famous for (among other things) its association with the noted nondualist philosopher Shankara. Kedarnath is, rather, a location where the sto-ries themselves suggest that there is something they cannot communicate about how this place is experienced. The story of the Pandavas invites us to approach through multiple modalities what Stella Kramrisch (1981) once famously called "the presence of Shiva."

2

Lord of Kedar

"Not far from Gaurikund, in between mountain peaks in the Mandakini val-ley, lies the Shri Kedarnath temple that is known as Kailasa, the original, eternal dwelling place of Bhagwan Shankar-Ji. There is no *linga* or image [Hindi: *murti*] here, there is only a high place [Hindi: *sthan*] in a three-sided shape[1] that is known as the back side of Shiva-Ji's buffalo." This excerpt from a modern pilgrim's guide points to the difficulty of expressing in precise language the exact textures and details of Shiva's form and presence in Kedarnath. Another example comes from the interview with Kedarnath *tirth purohit* Tiwari-Ji mentioned in the previous chapter. He said about the shape of the *linga*: "It is the formless shape of God [Hindi: Bhagwan]. . . . It was here from before, it is as if it comes from the ground, it is the real *linga*. . . . And [it is formless] so that people will not leave assuming that the *linga* has a particular form. . . .[2] *Ghee* is also important here. They [the visitors] are . . . laying on their hands [anointing the *linga* with *ghee*]." A third ex-ample arose in conversation with an elderly Kedarnath valley man who had spent numerous seasons in Kedarnath. He expressed the place/practice relationship in this way:

> Inside the temple there is a mountain. It is the Himalaya mountain inside the temple. And the God [Hindi: *Bhagwan*] there means the mountain. Whatever divine power [Hindi: *qudrat*] is there, that *qudrat* takes the form of God. But if you say to someone, "Worship a mountain," he won't do it, will he? "Worship the mountain," brother, no one will do it. What we old people did is we put the mountain inside the temple. We took that very mountain and built an impressive temple there, where now people come and exclaim, "*Vah vah,* what an enormous *linga!*" That's whose *puja* we do—we are doing the *puja* of the Himalaya. We do the *puja* of that lap of Himalaya in which we live.

Shiva's form and presence in Kedarnath, today, are marked with ambiguity. He is both formless and formed from the mountainous place, a rock in the ground. These accounts attest to this material ambiguity and connect it to one of the most memorable and distinctive pieces of modern ritual practice associated with Kedarnath: that everyone who enters the temple in the morning is allowed to massage ghee into the *linga* (a practice I discuss in more detail in chapter 4) with his or her own hands as an encounter with the formless, fluid materiality of Shiva's *linga* presence. Commercial visual culture in Kedarnath, about which I have written in detail, attests to this same ambiguous materiality (Whitmore 2012). What does it mean to say that Shiva is "in" the Himalaya, that he is "present" in a specific location?[3] What has it meant at different moments in the past? While anthropomorphic representations of Shiva are widespread, today it is fair to say that the *linga* is Shiva's primary form in our world (Bisschop 2009, 751). If, however, we think back to the second half of the first millennium CE in South Asia, it becomes a bit more difficult to envision exactly how Shiva's presence in the Himalaya was understood and experienced.

The recorded story of Shiva's ambiguously formed hyperpresence in the Himalayan landscape of Kedarnath began in the first millennium CE, when different Shaiva groups were imagining and engaging Shiva (and the earlier deity to which he was connected, Rudra) in diverse ways. The forms in which he was depicted and worshipped were during this period in considerable flux.[4] The second half of the first millennium CE in the Indian subcontinent was a time when numerous Shiva-oriented social formations, ways of depicting and imagining Shiva, and systems of Shiva-oriented ritual and philosophical thought were developing in dynamic and fluid ways. The increasing importance of Shaiva religious specialists and nascent institutional networks to politically astute kings influenced these developments. Much of this Shaiva world had begun to be present in the central Indian Himalaya and intersected in some way or other with Kedarnath.

Paying close attention to Shiva's ambiguity of form, and the multiple ways in which that form and presence were understood to connect to the Himalayan landscape, is important because these early understandings of form and presence functioned as one of the "initial conditions" of the complex eco-social system of Kedarnath. In their consideration of the application of the "science of chaos" to the practices of social anthropology, Frederick Damon and Mark Mosko (2005, 7) have noted that "sensitive dependence on initial conditions" is one of the hallmarks of chaotic phenomena. Tracing out these early patterns, while keeping in mind their continuously contingent and emergent character, is helpful for thinking about the present and recent past of this place. With this goal in mind, therefore, I review how early chapters in the history of Shaivism intersected with both Kedarnath specifically and the regions known today as Garhwal and Kumaon more broadly.

As Shiva-oriented myth, ritual practice, behavior, and philosophy moved into the second half of the first millennium CE and the second millennium CE, the surfeit of Shiva's powerful presence in the Himalaya began to mesh with the more general religious power of the Himalaya, multiple forms of the Goddess (some of the most important of whom are rivers), and the deity Bhairava. Multiple ways of understanding the nature and forms of Shiva, the Goddess, and Shaiva and Shakta (Goddess-related) deities and beings of power intersected in premodern constructions, imaginations, and experiences of Kedarnath. The place was not stamped by the worldview of a single group. Numerous relationships between deity, devotee, and place were imagined: relational, instrumental, devotional, mystical, alchemical, and nondual. My discussion of premodern Shaiva history of Kedarnath in a book that focuses on the twenty-first century is not meant to erase the historical distance between 600 CE and the late twentieth and early twenty-first centuries. Rather, I want to suggest that in the second half of the first millennium CE the overlapping of early understandings of Shiva's form with understandings of his presence in Kedarnath created a formative and foundational set of mythological, social, philosophical, symbolic, and material parameters.

SIXTH CENTURY CE

Let us imagine, then, what might have been evoked for someone living during the sixth century CE somewhere in the Indian subcontinent who had an interest in Shiva and heard about a place known as Kedara, associated with Shiva and located in the Himalaya. This was a time when much of what we would now call North India was in considerable flux. The vast Gupta Empire had begun to come apart, and newer and smaller rulers were stepping into its place. Because the Gupta emperors had officially been devotees and patrons of the god Vishnu, this meant that it made sense for newer and competing political formations to ally themselves with a different religious power, and Shiva in many places became the deity of choice (Bakker 2014, 4). The fifth and sixth centuries CE marked the beginning of what Alexis Sanderson (2009, 44) has termed "the Śaiva Age, " the age of the "dominance of Śaivism" (see also Fleming 2009b, 53). Hans Bakker (2014, 1–4, 10) and his team of researchers have recently shown that early manuscript versions of the Sanskrit text of the *Skandapurana*, a text that prominently describes Shiva's presence in the Himalaya and that generally wanted to "sanctify the landscape of northern India," both chronicled and helped shape the beginning of this new phase.

By this time a fierce deity called Rudra, a god with an uneasy relationship to many of the deities in the pantheon of divine beings who are described in and whose worship is enjoined in early Vedic texts, had come to be known as what Peter Bisschop (2009, 742) calls an "all-god"—the single being who is the source of all and to whom all devotion and religious attention must eventually be directed.

This Rudra came eventually to be known by names such as Mahadeva, Maheshvara, Sthanu, and Shiva. Shiva preferred to live in and around Mount Kailasa in the Himalaya as a solitary yogi and often expressed a preference for human and divine devotees who were inhabitants of the borderlands: non-Aryan local peoples and their non-Vedic local deities and supernatural beings, what Geoffrey Samuel (2008, 101–2) has termed local "guardian" and "protective" deities. Shiva was also beginning to display, somewhat paradoxically, both a personality committed to asceticism and solitude and at the same time an often erotic partnership with the daughter-in-law of the Himalaya: the goddess Parvati. As Benjamin Fleming (2009a, 444) has pointed out, this tension can be read in two different ways: as a synthesis in which over time the several different groups' portrayals of Shiva became fused in a single complex account and/or as a kind of mythological Moebius strip that offers insight into the ahistorical metaphysics of the god's deeper nature. Some groups at this time worshiped Shiva as an independent being, and others had begun to envision Shiva as a god who existed in special relationship to forms of the Goddess (Devi, Mahadevi, Parvati, *shakti*), who were themselves coming to occupy a more prominent place in those religious worlds that in the mid-first millennium CE were found in Puranic, epic, and agamic genres of Sanskrit texts.

In one of his most famous stories Shiva marries Parvati in her incarnation of Sati, the daughter of the great divine king Daksha, son of the creator god Brahma, who does not like his disreputable new son-in-law. Daksha does not invite him to the great fire sacrifice, even though he invites all the other gods. Sati (Parvati) becomes so incensed at this affront to her husband that in many versions of this story she kills herself. Shiva, upon hearing of the death of his wife, sends a personification of his anger (the deity Virbhadra, another early form of Bhairava) along with his fierce supernatural troops (Sanskrit: *gana*) to destroy the sacrifice of Daksha. Daksha ultimately becomes a devotee of Shiva after Shiva cuts off his head and replaces it with that of a goat. Shot through this story, and many other such mythological narratives of first-millennium CE South Asia, are complicated tensions between the Brahminical order (here represented by Daksha, the descendant of Brahma) and the diverse non-Brahmin human constituencies of Shiva. Shiva appears as the antinomian deity who lives beyond the geographical and social boundaries of the Brahmin-centric Vedic world and its affinity for the stability offered by Vishnu, and the place-based deities of his non-Brahmin constituencies often became his *ganas*. The putative location of Daksha's sacrifice, near the modern town of Kankhal at the lower edge of modern-day Garhwal, was already an important location for devotees of Shiva in approximately the sixth century CE (Bakker 2014, 173–81). Also woven into this competitive and diverse socioreligious milieu were numerous Buddhist and Jain communities.

Kedara-related names for Shaiva places were already multiple. Variations on the name Kedarnath that connected to a place or temple of Shiva already existed in many places in South Asia and as far away as Cambodia by the middle of the seventh century CE (Bisschop 2006, 35, 181; Fleming 2009b, 64–65). Peter Bisschop (2006, 14) has demonstrated, through his examination of early manuscript versions of the *Skandapurana* dating back to the ninth century, that possibly already between the sixth and ninth centuries CE Kedarnath (in that text referred to as Kedareshvara), by virtue of its presence on these *ayatana* lists, was known as a place of Shiva in the Himalaya. These early manuscripts of the *Skandapurana* place a great deal of emphasis on the Garhwal Himalaya.[5]

There are no conclusive references to Kedarnath as a Himalayan place that link conclusively to the Pandavas in early versions of the Sanskrit *Mahabharata*, arguably one of the oldest records of the established practice of pilgrimage in South Asia (Vassilkov 2002, 135). Kedarnath is, however, mentioned in connection with the Great Departure and eventual journey on the Great Path of the Pandavas into heaven in what Peter Bisschop (2006, 181) has assessed as "probably a relatively late interpolation," leading him to state, "We can conclude that Kedāra is absent in the MBh's [*Mahabharata*] lists of pilgrimage places, and thus the sanctity of Kedāra was recognized only after the normative redaction of the MBh." In the midst of a description of the Pandavas' Himalayan *yatra* that includes several other residences of Shiva that could arguably be located in the Garhwal Himalaya, part of the inter-polated *Mahabharata* passage relates the following about their visit to Kedarnath:

> After making reverence to Shiva [Sanskrit: Ishana] they bathed in the swan-water [Sanskrit: *hamsodake,* possibly some sort of special liquid]. Having seen the god of gods, Kedara, making an effort they touched him. After offering balls of food [Sanskrit: *pindam*] to their ancestors in the proper manner and making offerings [Sanskrit: *tarpya*] for their ancestors and for the gods, they drank water in the proper manner and went to the river Nanda. Then *he* [in my reading, Yudhishthira], having turned toward the Great Path, went to the place of snow.[6]

The fact that the Pandavas drink water in Kedarnath is notable. A particular connection with water is also one of the oldest things we know about the Himalayan Kedarnath. Early manuscripts of the *Skandapurana* introduce Kedarnath in this way just after describing the abode of Shiva at Mahalaya, an unidentified site that Bisschop notes may be suggestively but not conclusively connected to the modern shrine of Rudranath, which is today counted as one of the Panch Kedar, the system of five Shaiva temples in central Garhwal that includes Kedarnath. Here is the passage, in Bisschop's (2006, 66, for commentary, see 180–81) translation:

> More holy is Kedāra, which lies to the north of it, the supreme abode of Vṛṣāṅka. There Hara [Shiva] himself released the holy water from the mass of his matted hair. Men who drink that water become Gaṇas, Kuṣmāṇḍas, dear ones to Rudra. Brahmā,

Śakra, Viṣṇu, Soma and Kubera praise Parameśvara who is present there. They who take pleasure in *adharma* do not get to drink the water which streams from the head of Vibhu without Śiva's favour, but those who have drunk it are released from the fear of the chains of existence.

Kedarnath was one of the locations where it was believed that Ganga descended into the human world. Bisschop (2006, 182) further notes that "the liberating power ascribed to drinking the water at Kedāra is a recurrent theme in the Puranic descriptions of Kedāra." [7] By approximately the seventh century CE, Kedarnath in the Himalaya was known in the Tamil-speaking regions of South India as a mountainous place of water: an emphasis on water is found in a verse by the seventh-century Tamil poet Campantar: "on Kētāram, its peaks replete with water, where the unblinking celestials dwell." [8] Thus, in the earliest mentions of Kedarnath in the Himalaya, Shiva is present but not the only powerful entity of note. The efficacious place of Kedarnath also connects with the waters of the Ganga and the Himalaya more generally. The power of the place in this description is premised on the presence of Shiva but does not solely derive from him.

While the above *Skandapurana* passage refers to an anthropomorphic Shiva with "matted hair," it is important to realize that early Shiva-oriented groups worshipped and envisioned Shiva in more than one form, only some of which were anthropomorphic (Acharya 2009, 465; Bisschop 2009, 744, 746–47; Bakker 1997, 75–76). Benjamin Fleming (2009a, 444–45) artfully summarizes the complex situation in this way: "In the relevant materials dating from the second century BCE to the sixth century CE, there is no one dominant or consistent form. Nevertheless, one may see a general trend in the extant evidence: phallic imagery is gradually downplayed and abstracted over time." Different forms and combinations of Shiva were being created and worshipped at this time: an anthropomorphic figure with a club, an aniconic *linga*, a *linga* with an anthropomorphic figure emerging out of it, a *linga* with four or five faces (Srinivasan 1997, 260–81). Traces of this transformation in form and understanding can be seen in a famous story of Shiva that is often connected to Kedarnath. Vishnu and Brahma are arguing about who is more powerful. Suddenly a column of light appears. They decide that whoever can find the end of the column of light first is more powerful. The column of light is revealed to be unending, and Vishnu and Brahma must admit defeat. The column of light, it turns out, is Shiva in his *jyotirlinga* form, and this mythological narrative attests to both the rise of Shaivism and the transformation of understandings of the *linga* form during the multiple redactions of the *Puranas* that were under way throughout much of the first millennium CE.

This ambiguity of form was attested during the growth of a transregional network of sacred Shaiva pilgrimage places. Fleming (2007, 81–84) has shown, in part with specific reference to Kedarnath, that during the later redactions of the different parts and versions of the *Shiva Purana* (the Sanskrit text in which some of the most famous stories relating to Kedarnath and other Shaiva *tirthas* are found) it was not wholly clear exactly what was meant by the term *linga* or whether a *linga* was the only aniconic form taken by Shiva in the world. He has insightfully elucidated how the idea of the *jyotirlinga*, the *linga* made of light, represented a transformation of Vedic fire and earth worship into *linga* worship. Fleming has also demonstrated that the system of the twelve *jyotirlinga tirthas* knit preexist-ing somewhat separate Shaiva sites and various modes of worship in different parts of medieval South Asia into a single network of twelve places, the twelve *jyotirlinga*, that came to be understood as part of a single overarching Shaiva framework of sacred geography and religious practice.[9] This system naturalized the conflation of Shiva's various material forms (i.e., the forms through which he could be approached in ritual: *linga*, clay ball, fire, light, and different forms of earth) into the idea of the *jyotirlinga* system (Fleming 2007, 137, 167–69). Fleming concludes that this "cult of the *jyotirlingas*" arose between the tenth and thirteenth centuries (2).

EARLY SHAIVA GROUPS

Early Shaiva groups imagined a number of possible outcomes of their devotion to and focus on Shiva: for lay devotees material benefit in this life, residence in "the deity's paradise" after death, or a high-quality rebirth, and for the initiated non-lay practitioner "the attainment of the non-finite goal of liberation" (Sanderson 2013, 212). The "larger, lay Śaiva community" were sometimes termed "Followers of the Great Lord" ("Great Lord," in Sanskrit *Maheshvara*, is a common epithet for Shiva) (Bisschop 2010, 485; Bakker 1997, 70–79). Early "initiatory Śaivism" took two main paths, each with several branches: there were the systems of the "Outer Path" (Sanskrit: Atimarga) and the systems of the "Path of Mantras" (Sanskrit: Mantramarga) (Sanderson 2013, 212; Sanderson 1988, 664). One important early Atimarga Shaiva group, well attested by the fourth century CE, called themselves the Pashupatas (Acharya 2009, 464). The Pashupatas took their name from the epithet "Lord of Cattle/Beasts" (Sanskrit: Pashupati), one of the Vedic names of the deity Rudra that is often invoked in debates about the identity of the horned deity depicted on Indus valley seals and the connection of that deity to Shiva (Bisschop 2009, 741). The Pashupatas were Brahmin men who had renounced worldly life and envisioned themselves as the cattle of Rudra-Shiva. They engaged in extreme devotion to Rudra, physically imitated the actions of bulls, and recited Shiva-focused mantras. As Diwakar Acharya (2009, 459) puts it, "All initiates were ascetics, and all practice was aimed ultimately at liberation. Even though super-natural powers were achieved early on in their practice, a Pāśupata initiate was supposed to continue with his original quest for the end of suffering." This basic system in turn produced different Pashupata schools, one of which came to hold the influential view that the legendary figure of "the Lord with a Club" (Sanskrit:

Lakulisha), an "otherwise unnamed incarnation of Śiva," was the founder of the Pashupata system (Acharya 2009, 461). Lakulisha Pashupatas were also sometimes known as "Dark-Faced Ones" (Sanskrit: Kalamukha), a term that would become more important as time went on (Sanderson, 2013, 212).

The Mantramarga branches were also deeply involved in the formative early history of Shaiva traditions (Sanderson 1988, 664, 667). One Mantramarga group produced what would come to be known as the Shaiva Siddhanta system, associated with a set of Shiva-focused ritual and philosophical texts collectively called the *agama* and later more broadly with what would come to be termed tantric Shaivism. Their approach focused on teaching the initiated practitioner how to gradually, through the progression of a series of ritual and meditative stages, attain union with and/or transform into Shiva, the universal source of all, and thereby achieve "final liberation" (Davis 1991, 83). While today Shaiva Siddhanta traditions are thought of primarily as primarily South Indian, in the first chapter of the history of the Shaiva Siddhanta their denominations constituted a network that stretched across much of the subcontinent. They were adept at what Sanderson (2013, 213–14) has termed the "more visible" realms of religious practice of those on the Path of Mantras: administering temples and monasteries and strategically facilitating the ritual initiation of kings into the Shaiva community (see also T. Smith 2007). The Shaiva Siddhanta developed an important and influential overarching philosophical understanding of the nature of Shiva that was reflected in their iconography: that Shiva is both formed and formless (Davis 1991, 112–36).

By the ninth and tenth centuries some of the earlier networks of the Pashupatas had given way to those of the Shaiva Siddhantas. The Kalamukhas had also come to function as the inheritors of the Pashupata in many places across the subcontinent. By the ninth century they had already begun to be in positions of institutional authority in the region known today as Karnataka, and, while the historical record is not consistent, they appear to have focused on using transgressive behavior (ritual practice that involved interaction with impure substances, for example) as a way to move past the mundane distinctions of pure and impure (Sanderson 2013, 229–31; 1988, 666).[10] As I illustrate later in this chapter, it is reasonable to conclude that each of these (themselves not monolithic) groups had some direct or indirect connection to or knowledge of Kedarnath. We know that there were pilgrimages by Pashupatas and Kalamukhas to Kedarnath (Bisschop 2006, 182; Lorenzen 1972, 109, 173).

BHAIRAVA AND DEVI

The second stream of the Mantramarga, some survivors of the Atimarga, and a third, more Goddess-focused stream, the "Esoteric Path" (Sanskrit: Kulamarga), over time came to focus to a greater extent on two additional figures of divine power who were also coming to be connected prominently to Shiva in this mid-first-millennium period (Sanderson 2013, 212–13). The first was the Goddess (Devi, Mahadevi, Parvati, Durga, Kali, Shakti) in her multiple forms. Today Goddess-oriented Shakta traditions are deeply intertwined with Vaishnava (Vishnu-oriented) and Shaiva traditions. In many important ways this intertwining strengthened in the first half of the first millennium CE as Brahminical traditions began to fuse and incorporate non-Brahminical local and regional deities and supernatural beings, many of whom were female. This synthesis became further formalized and naturalized in ritual, philosophical, and textual contexts in the second half of the first millennium CE.

Goddess-oriented theisms are perhaps most famously attested in the *Devi Mahatmya* of the *Markandeya Purana*, which relates the victory of the warrior goddess Durga over demon foes whom the gods had been unable to vanquish. The forms of the Goddess most commonly associated with Shiva are Parvati (the god's marriage partner), Kali (one of the fierce manifestations of the anger of the Goddess), and *shakti*. In Puranic literature *shakti* can be presented either as a kind of energy, force, or anger or in personified form as a specific goddess. In personified form a *shakti* is presented both as an independent being and as the consort/partner of a god such as Vishnu or Shiva. The marriage of a local goddess with a god such as Vishnu or Shiva and the assimilation of the local goddess into the more universal figure of Devi (Mahadevi, the Goddess, Shakti, etc.) constitute enduring themes in Puranic literature and are indices through which the mixing of ethnic group, culture, geography, and language characteristics of South Asian religious traditions in the medieval and modern periods is often investigated by scholars (Chakrabarti 2001; Shulman 1980). All of this connects very directly to Garhwal: as the work of William Sax (1991) has chronicled, Devi, in her numerous regionally and locally important forms, is ubiquitously present and active in the modern Garhwali Himalayan landscape in myriad forms of local and regional significance.

The second important deity with whom Shiva came to be associated was Bhairava, who begins to be attested in the textual and archaeological record in the fifth to seventh centuries CE (White 2009, 485). Bhairava was a guardian deity, a demon lord who was known to inhabit cremation grounds and who came to be understood as the chief of Shiva's monstrous troops of followers (Sanskrit: *gana*). Bhairava was a manifestation of Shiva's anger and hence a form of Shiva himself, and he became the focal point for Shaiva systems of thought and practice that focused on transgression as a tool for religious progress. Bhairava was famously associated with a group known as the Kapalikas, or "someone who carries or deals with a skull or skulls (*kapāla*) on a regular basis" (Törzsök 2011, 355). The Kapalikas tended to model themselves after and worship Shiva in his Bhairava form. Their name comes from the narrative detail that when Shiva's angry Bhairava form cut off one of the god Brahma's five heads, the skull (Sanskrit: *kapala*) then stuck to his hand, making him a *kapalika*. Bhairava/Shiva, to cleanse himself of the pollution of Brahminicide (killing the god Brahma is like killing a human Brahmin priest), wandered from place to place for twelve years until he came to the city of Varanasi, where Brahma's decapitated head fell off and Shiva's transgression was erased, thus narratively establishing Varanasi as a center of Shaiva importance.

To digress into the Uttarakhandi present for a moment, it should be noted that Bhairava (more commonly known today as Bhairav or Bhairavnath) is one of the most important deities in in modern-day Garhwal. Along with forms of Devi, in various guises he functions as one of the important and ubiquitous bridges between the worlds of the *Puranas* and the local and regional worlds of regions such as Garhwal, in which his ubiquitous forms are appealed to for justice and revenge (Sax 2009, 27–28). Sax (2009, 25–50) has observed, in the context of a study on the cult of Bhairava among Garhwali Harijans, that Bhairav is a primary focus of local devotion and a far more active presence in the daily lives of most Garhwalis than Shiva. His identity in Garhwal centers on his ability to provide protection and justice. More generally, his character as the lord of spirits residing at the edge of demarcated times and spaces and guarding the transitions is continuous with his character in Nepal. I was once discussing Bhairav's relationship to Shiva with two local employees of the local hydropower plant that, in the recent years preceding 2013, powered the government facilities at Kedarnath. I had been feeling confused as to whether, in the contexts of Kedarnath and the Kedarnath valley, Bhairavnath was actually another form of Shiva or whether it made more sense to regard him as a lieutenant of sorts. The analogy we reached in the conversation that was satisfactory to all parties was that Shiva was the actual electric generator powered by the flowing water—full of vast amounts of power. Bhairavnath was the electric heater warming us as we spoke, a heater powered by that electric generator. The heater is an obviously smaller conduit for electricity, yet there is no direct way of accessing the electricity present in the generator without plugging in a smaller, external device.

Bhukund Bhairavnath is the form of Bhairavnath resident in Kedarnath.[11] His open-air shrine lies to the southeast of the Kedarnath temple at the lip of a small plateau that runs parallel to the main valley floor. Bhukund Bhairavnath is colloquially known as the lineage deity (Hindi: *kul-devta*) of the Kedarnath Tirth Purohit Association. The Kedarnath *tirth purohits* worship him with a *havan* and a *puja* several times each season, and these occasions are one of the most visible moments of the season, when the *tirth purohits* are in charge of the proceedings and the temple committee (the Samiti) is relegated to a secondary role. Indeed, one of the main local *havans* of 2007 serendipitously followed close on the heels of the day on which the Kedarnath Tirth Purohit Association received word that in their ongoing legal battle with the Samiti they had won an appeal at the Uttarakhand High Court in Nainital regarding their rights to collect ritual fees (Hindi: *dakshina*) in and around the temple. The shrine of Bhairavnath was open to the elements—every time the association asked him through his chosen human vehicle (Garhwali: *naur*) whether he would like them to build him a temple, he

FIGURE 6. The Bhukund Bhairavnath shrine overlooks Kedarnath.

refused. I also noticed, reinforcing Bhukund Bhairavnath's character in Kedarnath as a protector, that several of the important police officers in Kedarnath would make a point of paying their respects to Bhairavnath while standing in the temple courtyard during *shringar arati* at the time when the *pujari* would offer the *arati* flame in Bhairavnath's direction from the temple steps.

Bhairava and Devi, and often their myriad localized forms in the (Himalayan and other) landscape, became key foci for what have come to be termed systems of Shaiva Tantra. The term *Tantra* has a complicated history that is difficult to summarize. Geoffrey Samuel (2008, 9) has offered a useful general definition:

> a relatively coherent set of techniques and practices which appears in more or less complete form in Buddhist and Śaiva texts in the ninth and tenth centuries CE. This comprises a number of elements: elaborate deity visualisations, in which the practitioner identifies with a divine figure at the centre of a *maṇḍala* or geometrical array of deities; fierce male and particularly female deities; the use of transgressive "*Kāpālika*"-style practices associated with cremation grounds and polluting substances linked to sex and death; and internal yogic practices, including sexual techniques, which are intended to achieve health and long life as well as liberating insight.

The Himalayan landscape in and around Kedarnath came to be understood as a potent resource for powering such tantric techniques, as several passages from the Sanskrit texts of the *Kedarakhanda* and the *Kedarakalpa* discussed later in this chapter attest.

TRANSREGIONAL CONNECTIONS

A famous story in the *Mahabharata* constructs the Himalaya as a place where the encounter with Shiva functions as the bridge between the North Indian plains and the non-Aryan worlds of the Himalaya. Arjuna, one of the Pandava princes whose saga drives the plot of the *Mahabharata*, journeyed to the Himalaya during the period of the Pandavas' exile in the forest. There Arjuna fought with a disguised Shiva, thinking him to be an indigenous local hunter, a *kirata*. Shiva revealed himself to a suddenly awestruck and devoted Arjuna and granted him an invincible weapon suggestively named "the weapon of the Lord of Beasts" (Sanskrit: Pashupatastra).[12] The Garhwal Himalaya is one region where this encounter has been understood to take place, and it makes a degree of sociohistorical sense. By the middle of the first millennium CE, Garhwal had begun to function as both periphery (the edge of "the Land of Noble Ones," Sanskrit: *aryavarta*) where indigenous peoples lived who might possibly have migrated earlier from Central Asia, and center (the pure, remote Himalaya, home of sages, Shiva's home of Kailasa, and north of that Mount Meru). Kedarnath was one of the places that served to mark the boundary of the *aryavarta*, whose boundaries a powerful king might imagine that he could bring under his rule under optimal conditions (Sircar 1971, 11).

Peoples at the time indigenous to Garhwal and Kumaon had already been interacting with many of the early Shaiva groups who had been coming from the plains for centuries, particularly Pashupata groups, and some regional rulers were Shiva devotees or supporters of Shaiva groups, particularly Lakulisha Pashupatas (M. Joshi 2007, 216–20; Ḍabarāl, n.d., 170–73). The Shiva temple of Gopeshwar in the Chamoli district of Garhwal is particularly important for charting this pattern of early connections to Pashupata Shaivism in Garhwal and Kumaon (M. Joshi 2007, 224, 227). Much of the social, ritual, and philosophical diversity found in the early Shaiva worlds across the subcontinent was also present in the Garhwal and Kumaon regions of the Himalaya.[13] It is reasonable to assume that Kedarnath would have been one focal point for all of this, and that, along with other important old Shaiva sites in the region such as Jageshwar, Kankhal, Gopeshwar, Pandukeshwar, and Lakhmandal, it may have functioned as one of the nodes through which Shaiva groups from outside what we now call the central Indian Himalaya became connected to the region and interacted with each other. The work of art historian Nachiket Chanchani (2015, 2014, 2012) on the presence of Shaivas in Garhwal and Kumaon offers useful material support for this point.

Many accounts of Garhwali history, including those sometimes told at Kedarnath, correlate the Brahminization of the region and the establishment of monastic linkages to Badrinath and Kedarnath with the figure of the famous philosopher Shankara.[14] Brahminization, or (to use the term initially developed by anthropologist M. N. Srinivas to describe what he observed among the Coorgs, a South Indian tribal group) "Sanskritization," refers in a general sense to "a process by which the group gradually adopted characteristics of surrounding 'Sanskritic' culture in order to raise its status in relation to those cultures" (Srinivas 1952; Samuel 2008, 78).[15] This process often involved a transition to a social framework that valued the knowledge, status, and authority of Brahmins and valorized Sanskrit. While the historical realities are considerably more complicated, the visit of Shankara is popularly understood to have turned the character of the region in the direction of Shaivism (and to lesser extent, Vaishnavism) and away from the influence of Vajrayana and Mahayana Buddhism (Jain 1995, 21, 53). Shankara was one of the most famous and influential thinkers in the history of the subcontinent and in many circles was and is regarded as an avatar of Shiva. In addition to the profound influence on the development of (nondual) philosophical thought in South Asia ascribed to him, he is traditionally credited with the establishment of the Dashanami monastic order, one of the primary institutions that structure the lives, practices, and social worlds of Shaiva renunciants through to the present day. According to traditional sources, Kedarnath is one of the places where Shankara may have passed away (Clark 2006, 155). Govind Chandra Pande (1994, 52) places the dates for Shankara's life between 650 and 775 CE.[16]

The journey of Shankara into the Himalaya attests to the fact that entry into the "Shaiva Age" coincided with a rise in the popularity of *yatra* and the continued incorporation of peripheral areas like Garhwal and Kumaon into the emerging sacred geographies of the Puranic imagination, geographies that would be enacted through the practice of *yatra* (Nath 2009, 202–5). The practices that Puranic texts associate with *yatra* to specific *tirthas* often operated as syntheses of preexisting Vedic and Brahminical frameworks with other popular local and regional traditions. These practices included performance of the following at *tirthas*: "fire oblations . . . meditation . . . satiation of gods, sages and ancestors . . . ancestral rites . . . austerities . . . religious-gift-making," *puja*, listening to stories about *tirthas* to gain merit, *yatra* to the *tirtha*, circumambulation, shaving of the head, chanting and reciting of verses many times, and on occasion "religious suicide" (Nath 2001, 131–32). This list describes captures many of the practices found at Hindu pilgrimage places today.

By the tenth to thirteenth centuries Kedarnath was already well established as a place of pilgrimage and a site of royal patronage (V. Pathak 1987, 92–93; M. Sharma 2009, 53; Ḍabarāl, n.d., 478–85; Naithānī 2006, 179–82). Kedarnath had become part of the network of *tirthas* linked by what Diana Eck (2012, 39) has called a "complex grammar of sanctification" that spread throughout the subcontinent, a web of paths and places linked by devotion, politics, custom, and divine power. Kings were traveling to Kedarnath, and thousands of kilometers to the south poet-devotees such as Campantar had already been singing their devotion since the seventh century and mentioning Kedarnath in the Himalaya. Local rulers in the tenth to thirteenth centuries were sponsoring temples that incorporated their kingdoms into larger-scale sacred geographies that, according to Chanchani (2015, 36), enacted "an evolving idea of India . . . a cohesive geo-cultural entity extending from the shores of the Indian Ocean up to the snow-capped Himalayan peaks" as a way to strengthen their rule (see also M. Sharma 2009, 21–22). Rulers forged these connections by supporting temple construction and maintenance in ways that drew artisans from outside the region, by taxing trade and agriculture, by carrying out pilgrimages that linked them to the Gangetic plains, and by using land grants to lure Brahmins from the plains. Chanchani (2015, 35) shows that this pattern may be fruitfully used to understand certain aspects of the situation in Garhwal and Kumaon and suggests that it can be applied to understanding the development of the famous shrines of Kedarnath and Badrinath.

NATHS AND VIRASHAIVAS

Two additional Shaiva communities became significant in Garhwal between the thirteenth century and the beginning of the nineteenth century, both of which displayed a distinctive understanding of the textures of the place-deity-devotee relationship. By the sixteenth century, the Naths had a strong presence.[17] David White (1996, 97) has noted connections between the Nath-Siddhas and the Pashupatas and Kapalikas. One of the important differences, however, in the practices and doctrine of the Naths was the internal visualization of the often externally transgressive practices associated with the Pashupatas and Kapalikas rather than actual external physical performance of those practices. Naths also differed from earlier Shaiva groups by orienting their tradition around a series of gurus, beginning with the primordial guru Shiva and thence the legendary human Matsyendranath (Macchandar) and his disciple Gorakhnath (Mallinson 2009; White 2015). For Naths, Kedarnath and Gaurikund are significant as a set of places where the union of Shiva and the Goddess and their powerful bodily fluids are especially accessible (White 1996, 196, 245–46). The material power of the ground and water could serve as fuel for the transformation of what David White (1996) has termed "the alchemical body." The Sanskrit text *Kedarakhanda* (to be distinguished, as we will see, from the *Kedarakhanda* section of the *Skandapurana*) places an ashram of Gorakhnath near Gaurikund, and King Ajay Pal was connected to the Nath tradition and to the Nath yogi Satyanath and gave him land, an event that can be approximately dated to the sixteenth century (Ḍabarāl, n.d., 106, 182–83; White 1996, 112 and 407–8n177; Rawat 2002, 34; Negi 2001, 17). Today there is a dense network of Nath shrines and sites of importance for Nath yogis in Garhwal and Kumaon (Kukaretī 1986).

The second of these more recently prominent Shaiva groups is the Virashaivas. They are the *pujaris* of Kedarnath today and assert that they have held this position for millennia since having received the region of Ukhimath (originally termed by them "Ushamatha" in memory of a marriage between Usha, daughter of Shiva's demon devotee Banasur, and Krishna's grandson Aniruddha) as a gift from Janamjeya, the great-grandson of the Pandava prince Arjuna (Hiremath 2006, 12, 28–29).[18] According to the versions of Virashaiva tradition reported in the modern-day Kedarnath valley, the Virashaiva teacher Ekorama, one of five primordial incarnations who first emerged from each of the five faces of Shiva's linga, founded Kedarnath.[19] Kedarnath is described as one of the five headquarters (Hindi: *math*, or *peeth*) of what Virashaivas in the Kedarnath valley refer to as the Virashaiva tradition, along with Rambhapuri, Ujjain, Shri Shailam, and Varanasi. It is known as the Vairagya Peetha today (Hiremath 2006, 3). Each *math* is led by a *jagatguru*. Bheemashankaralinga Shivacharya Bhagavatpadaru is the current Kedarnath *jagatguru* and the *rawal* of Kedarnath. His students (Hindi: *cela*) serve as the *pujaris* for Kedarnath, the Vishvanath temple in Guptkashi, the Madmaheshwar temple in the Madmaheshwar valley, and the Omkareshwar temple in Ukhimath.[20]

It is known that the earliest versions of the diverse groups today termed Virashaivas, who tend to be connected to regions today known as the states of Karnataka and Maharashtra, were found in many of the same places as the Kalamukhas, but the relationships of these groups are not wholly clear (Lorenzen 1988; Michael 1983; Leslie 1998; T. Smith 2007, 305). Recent scholarship about the history of the Virashaivas suggests that the idea of a "Virashaiva" identity, today often synonymous with or connected to the label "Lingayat," may have solidified only in the fifteenth and sixteenth centuries (Shobhi 2005, 8, 252–59, 269, 295–96). The early recorded history of this group involves a complex relationship to other Shaiva groups. This set of traditions displayed a different relationship to Shaiva places than the Naths: many of the early voices in this set of traditions contrasted the static (*sthavara*) nature of place-based temple-centric ritual with the moving (*jangama*) and fluid experiential world of the embodied, wandering Shiva devotee devoted to a personal and portable *linga* (Ben-Herut 2016, 2015; Leslie 1998; Lorenzen 1988; Michael 1983).[21] The goal of the devotee was the ultimate experience of a nondual Shaiva reality in which devotee and deity both merged and transcended the binary imposed by the sense of devotee-deity relationship.

PREMODERN SHAIVISMS IN GARHWAL AND KUMAON

While there are many importance differences among the worldviews and practices of these different Shaiva groups, their common ground should also be emphasized. Keeping this idea of common ground in mind, I think that much can be understood about earlier relationships of the groups sometimes known today as Virashaivas to Kedarnath through comparison with the social history of another *jyotirlinga*: Shri Shailam in Andhra Pradesh. As Prabhavati Reddy (2014, 78–81) has documented, Sri Shailam served between the seventh and sixteenth centuries as a point of meeting and mingling among local, regional and transregional Shaiva traditions that included the Pashupata, Lakula, Kalamukha, Lingayat, Kapalika, southern Shaiva Siddhanta, and Smarta Shaiva traditions (see also White 1996, 110). Something similar may have happened in Kedarnath.

It is clear that by the beginning of the nineteenth century Kedarnath was of significant concern for multiple lay and monastic Shaiva groups and a focus of the more general interest of Shiva-oriented *yatris*, kings, and renunciants from across the subcontinent who were engaged in practices of pilgrimage, devotion, rule, and patronage. It is also clear that there were myriad overlapping understandings and experiences of the presence, form, and nature of the Shiva felt to be present in Kedarnath. There were different, overlapping models for thinking through the relationship of the Shiva devotee to Shiva that could connect to Kedarnath: a relationship that would deepen infinitely, a relationship that would turn into mystical union, a relationship of infinite devotion, a relationship in which the devotee would become the god, a relationship that would produce power that could be used for material benefit, a relationship that would enable liberation from the prison of mundane existence, and a relationship that could recalibrate social relationships and political contestations for power and authority. Foundational stories and important ritual practices connected to Kedarnath, as we will now see, resonated across this entire range.

THE STORYING OF KEDARNATH

Some of the most famous Kedarnath-related stories, part of the broader web of stories that frame places like Kedarnath, are found in the *Shiva Purana*, the best-known and most commonly referenced Sanskrit text related to Kedarnath. Both in the Kedarnath valley and more generally, the *Shiva Purana* is one of the Puranic texts that is commonly recited and expounded in the context of multiday recitations (Hindi: *katha*). Recitations of the *Shiva Purana* often occurred in Kedarnath during my time there: in 2007 three were scheduled and two were performed. One of those performed was a very public recitation sponsored by the Samiti, and the other was a private recitation sponsored by a group of *yatris* who brought their own reciter with them and who resided in Kedarnath for the duration of the *katha*. Texts of the *Shiva Purana* are available for purchase in the bazaar. Also present in the Kedarnath bazaar in 2007, but to the best of my knowledge in only two shops, was the previously mentioned tantric text of the *Kedarakalpa*. There is also a third text of critical importance for understanding Kedarnath: the *Kedarakhanda*. This text claims to be the first section of the previously mentioned *Skandapurana* (which bears the same title) but in reality differs in content, focusing almost completely on providing a description of the sacred geography of Garhwal. A discussion of key passages in these three texts shows how Kedarnath functions as a place that (again remembering Edward Casey's notion of place as a kind of intersubjective gathering) *gathers* and integrates the different Shaiva understandings of Shiva's present in the Himalayan landscape and how devotees might engage the power of the deity/place that has been the subject of the chapter thus far.

THE *SHIVA PURANA*

In one of the *Shiva Purana*'s most substantive passages mentioning Kedarnath we meet a dual incarnation of Vishnu in the forms of the sages Nara and Narayana, who are carrying out penance and ascetic practice.[22] More specifically, they are worshipping Shiva. After a long time Shiva, pleased with their devotion, offers them their choice of boon. The two sages reply: "O lord of gods, if you are delighted, if the boon is to be granted by you, O Śiva, stay here in your own form [Sanskrit: *svena rupena*] and accept the devotion of your devotees." As the passage continues, we find out that the setting of this episode is "Kedāra": "Thus requested, lord Śiva himself stayed in Kedāra on the Himavat in the form of Jyotirliṅga. He was worshipped by them for helping the worlds and for appearing in the presence of the devotees." A bit later in the same passage the Pandava-related story appears:

> It was he who on seeing the Pāṇḍavas assumed the form of a buffalo, having recourse to his magical skill and began to run away. When he was caught by the Pāṇḍavas he

stood with his face bent down. They held his tail and implored him again and again. He remained in that form in the name of Bhaktavatsala. His head portion went and remained fixed in the city of Nayapāla [Nepal]. The lord stood in that form there. He asked them to worship him in that trunkless form. Worshipped by them, Śiva remained there and granted boons. The Pāṇḍavas went away with joy after worshipping him. After obtaining what they desired in their minds, they were rid of all their miseries.

There in the shrine of Kedāra, Śiva is directly worshipped by the Indian people. He who makes a gift of a ring or a bracelet after going there becomes a beloved of Śiva. He is endowed with the form of Śiva. On seeing that form of Śiva, a person gets rid of sins. By going to Badarī forest he becomes a living liberated soul. On seeing the forms of Nara, Nārāyaṇa, and Kedāreśvara, undoubtedly he can achieve liberation. The devotees of Kedāreśa who die on the way are released from rebirth. No doubt need be entertained in this respect. Going there, with pleasure, worshipping Kedāreśa and drinking the water there, a person is released from rebirth. O Brahmins, in this Bhārata country people should worship with devotion Nara-Nārayaṇeśvara and Kedāreśa. Although he is the lord of the universe still he is particularly the lord of Bhārata. There is no doubt that Śiva Kedāra is the bestower of all desires. O excellent sages, I have narrated to you what you have asked for. On hearing this narrative the sins disappear at once. No doubt need be entertained in this regard.

Different ways of talking about Shiva's form coexist in this passage. It is possible to see a tension here between the idea that Shiva is hyperpresent in his Himalayan abode and the well-documented character trait that Shiva is a god, as Don Handelman and David Shulman (2004, 220) have put it, whose "body is itself a dense mass of joyfulness" but whose "habit" is to "slip away." This story fixes God's desire to slip away in the soil of the place itself. True *darshan* of Shiva is, to extend the interpretive path of Handelman and Shulman, *darshan* of his desire to be less than fully present and at the same time, of his dynamic flow both utterly into and away from what is happening. In one sense this is why he initially distances himself from the Pandavas—to bring them to the place where his absence is most present and powerful. These passages suggest that Shiva is both eminently available, graciously manifesting at the request of the sages Nara and Narayana, and at the same time reticent, liable to run away and in need of persistent supplication. This passage, in what is now evidently a persistent theme, also emphasizes the importance of drinking water in Kedarnath. I often heard this verse quoted in Kedarnath in 2007 and 2008: "When one has drunk that Kedar-water that transcends *samsara* and cuts the net of *papa*, there is no rebirth. One will not be bound in a womb but go to the eternal state" (Lakṣmīdhara 1942, 8:230).[23]

The importance of the Himalaya also hovers in the background of this story. Amid many of the cultural geographies of the natural world in South Asia, the Himalaya has stood as a region apart. It has been the zone of purificatory power to which ascetics retreat to meditate, following in the footsteps of the famous sages of the past, as well as a set of culture areas with different local deities more connected to Tibet and Nepal than to the Gangetic plain and the rest of the Indian subcontinent. It is the source of the Ganga and Yamuna rivers. The Himalayan mountain Kailasa, famed as Shiva's most preferred place of residence, has been understood by many strands of Hindu, Buddhist, and Jain traditions as Mount Meru, center of the world (Eck 2012, 199), and is itself a famous pilgrimage destination. Specifically, within the borders of *kedaramandala,* the karmic fruits of meritorious action are multiplied. As William Sax (2009, 53) has observed, the Garhwal region of Uttarakhand demonstrates an especial density of "world famous pilgrimage places" that are mentioned in different *Puranas* and versions of the *Mahabharata.* The Kedarnath valley is replete with references to ancient sages such as Agastya (in the village of Agastmuni) and Jamadagni (in the village of Rabi just above and west of Fata). There is a public debate every year in the village of Kavilta (located behind Kalimath) about whether Kavilta is the natal place of the famous Sanskrit playwright Kalidasa. Kalidasa gave the Himalaya pride of place in one of the most famous dramatic works of Sanskrit literature, *The Birth of Kumara:*

> There is in the north
> the king of mountains,
> divine in nature, Himálaya by name,
> the abode of snow.
> Reaching down
> to both the eastern
> and the western oceans,
> he stands
> like a rod to measure the earth.
> (Kālidāsa 2005, 25)

In the worlds of Puranic narrative the Himalaya is both geology and character—Shiva's preferred residence and his father-in-law. Arjuna meets Shiva in the Himalaya in the guise of a tribal hunter and, after being humbled in a contest with him, becomes his devotee and is gifted with the Pashupatastra weapon in return. The Himalaya is the location for some of the most important stories that provide insight into Shiva's nature. It is in the *daruvana,* the Himalayan forest of pines, that Shiva decides to teach a harsh lesson to a group of sages who have been ignoring him and mindlessly overfocusing on arduous ascetic practices. Vishnu, in his beguiling female form of Mohini, distracts the sages while Shiva goes to see the wives of the sages in their village. Upon his arrival they all fall straightaway into uncontrollable desire. The sages, realizing what has happened, use their sacrificial fire and generate a series of world-ending weapons with which they attack Shiva, culminating with (in the fourteenth-century *Kanta-puranam* version of this story) the production of the destructive being Muyalakan. Shiva then begins to dance his *tandava* dance on the head of Muyalakan, and the power of the sages is destroyed, along with the egocentric karma that created the situation (Handelman and Shulman 2004, 4–14). This image, of Shiva who is the Lord of Dance (Sanskrit: Nataraja), becomes one of the famous modalities of Shiva's presence in the world.

The Himalaya is also the backdrop for many of the central episodes in the relationship of Shiva and Parvati, some of which are held to have happened near Kedarnath. Just before the trailhead at Gaurikund one may stop at the temple of Munkatiya, where Shiva cut off the head of Ganesha as he tried to prevent Shiva from seeing Parvati before she was ready after her bath. Gaurikund, as Diana Eck (2012, 226) has pointed out, is according to some where Parvati, also known as Gauri, performed the "severe spiritual austerities" necessary to convince the ascetic god that he should marry her. With the help of Kama, the god of desire, Parvati finally agreed and their union produced a son, Skanda, who was able to kill the otherwise invincible demon Taraka. These stories establish the complexity of Shiva's identity as both ascetic and family man and ground that complexity in a kinship relation with the Goddess and the mountains.

THE *KEDARAKHANDA*

There are also region-specific texts and story cycles that localize these Puranic stories in specific regions. The *Kedarakhanda* is one such text— a medieval/early modern Sanskrit text that focuses on the religious geography of the region known today as Garhwal.[24] When one examines the *Kedarakhanda* with a Kedar-centric eye, it becomes apparent that Kedarnath is effectively a focusing node for the broader power of the region. The first story about Kedarnath in the *Kedarakhanda* immediately follows an account of the descent of the Ganga and the elucidation of her multiple watery flows.

> Vasishta said:[25]
> Having heard this excellent story, Parvati, filled with devotion, said to her husband Shiva, "God of Gods, ruler of the world, you are he whose grace is the final objective of devotees. You have now told me about the ten streams of the Ganga. Abode of all that which is, tell me the names and greatness of the country where those flows went. Tell me their greatness in detail; I want to hear this now. Where did those Brahma-arisen waters meet? Tell more about the greatness of those places. I am your devotee, Lord of Gods, you alone are dearer to me than breath. People do not conceal from those who love them. Lord, you alone are the maker of the world, the upholder of the world. By you alone, Lord of Gods, are these three worlds permeated entirely. Tell in detail in which countries are the ten flows of the Ganga."

Parvati's questions are similar in form (if not fully in import, considering the metaphysical underpinnings of a dialogue between Shiva and Parvati/Shakti), to those that *yatris* might ask their *tirth purohit*. They also remind the reader about Shiva's complicated identity as both ascetic and husband.

The Lord said:

Listen, Devi, best of women, matchless one—I will answer what you have asked, and it has never been told to anyone else before. Effort should be made to keep this marvelous and wonderful secret hidden. By the mere sight [of such places] millions of *papa*-generating actions are burnt up as if by fire, ruler of the ruler of the gods. Hear, Devi, how once the god-sired Pandavas, the famous and mighty sons of great-souled Pandu, were disgraced by their killing of their own lineage and afflicted by their murder of their own guru. Having killed Drona and all the others, they were greatly troubled, and with their hearts burning and greatly agitated with grief, with all their deeds come to foul ends, they sought refuge with Vyasa. Possessing minds into which impurity had already entered, they said to great-souled Vyasa: "Blessed Vyasa, all of us have come for refuge with you. How can we, Brahman, whose selves are now full of wickedness [Sanskrit: *papa*], attain liberation? We are disgraced by the killing of our own lineage and afflicted by the murder of our guru. You are our only refuge; give us a clear command. Brahmin, by what action can we attain the highest state? Having taken pity on us because we are your descendants, tell us this."

Vyasa said:

"You Pandavas who are killers of your lineage, listen! What is generally valid as a cure for all cases does not hold for those who have killed members of their lineage unless they go to the abode [Sanskrit: *bhavanam*] of Kedar. Go there! That is where Brahma and other gods who are desirous of the *darshan* of Shiva stay established in purificatory ascetic activity [Sanskrit: *tapas*], purified of their actions as they carry out the highest *tapas*. That is where Ganga is, the most eminent and chief of the numerous rivers. That is where the Lord Shiva lives, along with his numerous mighty followers and kings. That is where the gods, along with the *gandharvas* and *yakshas* and *rakshasas* and bulls, have their sport every sunset of one half of the month of the scorpion. Many instruments sound there and the chanting of the Vedas is heard. That is the place where those who are dead become Shiva, without doubt. Who can describe the greatness of that area [Sanskrit: *kshetrasya*]? He who has given himself over to numerous *tirthas* and remembered that area will get liberation. Go to that divine place [Sanskrit: *tridashasthanam*] known as the Great Path. That is the best place for expiation of the *papa* of killing Brahmins."

Of distinct importance in this paragraph, and throughout many parts of the *Kedarakhanda,* is that there is a good degree of semantic overlap between the region of Kedara (Sanskrit: *kedaramandala*), roughly conforming to modern Garhwal, and the area of Kedarnath proper. In this small section alone one finds a number of different place words that set up this overlap: *bhavana, sthana, kshetra,* and later on *sthala.* This variation of different location-terms is typical in passages from the *Kedarakhanda* relating to Kedarnath. We also see that the story dwells on the weight of the *papa* carried by Draupadi and the Pandavas. Further, the passage clearly functions as an example of the Puranic mechanisms of localization and incorporation: semidivine (Sanskrit: *gandharva, yaksha*) and demonic (Sanskrit: *rakshasa*) beings, many of whom may have been preexisting local deities with connections to the natural world or to a specific locale, are here incorporated into Shiva's world as his followers.[26]

> The Lord Shiva said:
> "Thus Vyasa spoke. Having heard, the Pandavas delightedly reverenced him, circumambulated him, and went to Mount Kailasa. They lived there and attained the highest condition. The *tirthas* by whose selfless worship [Sanskrit: *seva*] those mighty ones became pure, the attainment of that place is difficult even for the gods. It is fifty *yojanas* [Sanskrit: a Vedic unit of measurement] by thirty *yojanas,* and this place is whence one goes to heaven—that mighty place is not earthly. From the border of Gangadvara to the white mountains and from the banks of the Tamasa to the underside of Boddhacala is known as the auspicious region of Kedara, a raised place [Sanskrit: *sthala*] separate from the world."

This first message makes no specific reference to Kedarnath but rather delineates the region as a whole. This recalls my stipulation in the Introduction that we may productively view Kedarnath as a complex eco-social system with fuzzy edges.

The second story (Nautiyal 1994, 144–47) involves a development whereby living creatures transform to Shaiva beings because of their proximity to Kedara, a phenomenon described in the *Puranas* as "the state of having the same form as" (Sanskrit: *sarupata*) a specific deity.[27] A hunter in pursuit of quarry finds himself in a forest that lies, unbeknownst to him, at the edge of the Kedara region. An inexplicable darkening of the sun has just caused him to narrowly miss shooting a golden deer (who, he did not know at the time, was the sage Narada). He sees a snake eating a frog and approaches to investigate. "Just as he went closer, he saw a frog in a hole being quickly swallowed by a large snake. Just as the black snake swallowed this frog, the frog turned into a noble trident-bearing entity wearing a serpent as a sacred thread, with resplendent locks, bearing a half-moon, glorious like Kailasa, dancing, adorned by his followers, blue throated, wearing the skin of an elephant." As the passage proceeds, the hunter encounters another wondrous sight that challenges his mental faculties even more: the apparently spontaneous transformation of a tiger killing a deer. The deer becomes Shiva and the tiger, killed by another hunter, becomes a bull on which Shiva then seats himself. The hunter is rescued from mental breakdown by meeting the sage Narada and learning that these transformations are happening because he is in the neighborhood of Kedara. As living beings cross the event-horizon of the *tirtha*/region, they achieve the *sarupata* of Shiva.

This theme of the fluidity of identity and form brought on by arrival/residence in the Kedara region/*tirtha* may be found in many places in the text. The *Kedarakhanda* tells of a merchant couple who adopt a river and treat the river as a daughter, or a merchant who takes a mountain as his son when he is unable to become a father through normal means (Nautiyal 1994, 787, 779). One reads about beings and natural entities inhabiting the region who were previously wicked, from a lower social class (Sanskrit: *varna*), or a demon (Sanskrit: *rakshasa*) or about pious Brahma-*rakshasa* who have taken up residence in the region (sometimes in groups of ten million) until they may achieve final liberation (Nautiyal 1994, 163, 542). One river was once a low-caste human (Nautiyal 1994, 700–702).

These stories paint a picture of the Kedara region as a place whose extraordinary purificatory power creates a fluid continuum among different kinds of people, demons, the natural environment, and deities rather than a sharp distinction among different classes of entities. This power both depends on the presence of Shiva and is independent of that power and connected to the Himalayan region itself. *Yatris* in Kedarnath often told me that that the results of meritorious action performed on *yatra* would be exponentially multiplied by the power of Kedarnath or of the region. A sense runs through these accounts of the Himalaya as a region that is somehow in motion and whose form can change. This, of course, resonates with the fact that the Himalaya are young, geologically and seismically active. The mountains *can* change. Also emergent is the theme that this region is somehow cut off or separated from the North Indian plains, not just in geographic but also in sociocultural and even existential terms. Life is different in the mountains. Travel is difficult, and the mountainous landscape can impose its own rules. This theme runs through to present-day understandings of what it means to be *pahari*, a mountain dweller as opposed to someone from the plains.

The geographic survey of the Kedarnath valley in this part of the *Kedarakhanda* text eventually comes around to what are known as the Panch Kedar, the five Kedars (of which Kedarnath is the most famous). Toward the end of this section the narrative returns to Kedarnath and the Pandavas, with Shiva as the narrator.

"Listen, Devi, best of women, Parvati, to my *tirthas*. There the excellent Vaitarani flows, who causes the crossing over of ancestors.[28] There by the act of offering food to the ancestors [Sanskrit: *pindadana*] one gets the fruit of a hundred thousand Gayas. There is a pleasing Shiva-face there bearing all sorts of adornments. By its *darshan* people are liberated. Formerly, I, the ruler of the gods, was fiercely and persistently sought by the Pandavas, all connected to the killing of members of their lineage, for the purposes of purification of their *papa*. Seeing them, I went to the region of Kedar. They followed behind me for a very long way. Seeing them coming close, I entered the ground. Seeing me near, they came behind me, and in the location known as Kedar, Goddess, they touched my auspicious back. By that mere touch they were liberated from the murder of their lineage, and my back part, Parvati, is established there even today. The lord of Kedara is reported in all the three worlds as liberation-bestowing." (*Kedarakhanda* 52.1–8, ed. Nautiyal 1994, 177–78)

This passage, a different version of which we have already seen as part of the narration offered by Tiwari-Ji and as part of a passage in the *Shiva Purana*, indexes the most famous and most-often told stories about Kedarnath that one hears today. In it, the persistence of the Pandavas is what causes Shiva to dive into the ground, thus activating his specific, hyperefficacious presence in Kedarnath. Yet even as we read this we recall that Shiva is telling this story to Parvati as part of a longer description about Garhwal in which Shiva effectively functions both as pilgrim guide and as the ultimate object of the aspirations of the devotee who journeys into the region. The relationship of Shiva and Parvati and their presence in the Himalaya is much deeper, broader, and infinitely longer than this brief account, as most readers would know. It is part of the play of conjoined Shiva-Shakti to listen to stories about their own true natures. Shiva's presence in Kedarnath is both historicized and ahistorical. This version also enshrines an important feature of the place—it is a place defined by the mark (Sanskrit: *linga*) of Shiva's present absence, a shape that tells us that the hyperpresent god is on his way to somewhere else. This story, this *myth*, tells us that grasping Shiva's presence in this place is difficult; it does not allow us to separate out stories of place from narratives of self. These stories, I think, rhetorically self-signal the limits of what can be expressed about the nature of the god and the potential for self-transformation in the place.

The second half of the *Kedarakhanda* focuses on a site-by-site elucidation, with digressions, of important shrines and natural features in the area. The geographically localized bulk of the text discusses the region of Garhwal in detail. Here we see charted a tantric geography in which the ability of many different locations to generate supernatural powers (Sanskrit: *siddhi*) and transform matter is exactingly detailed. Sites connected to the Kedarnath area in this regard include Retas Kund, the pool of quicksilver/semen in Kedarnath (Sanskrit: *retasa*, of alchemical importance to Nath-Siddhas, as noted by David White), red-colored water that turns metals to gold, and a site near the ashram of Goraksha where one can attain the ability to create pearls by touch (White 1996, 246). The presence of Bhairava and Kali are noted, as is the especial efficacy of the area for the funerary rituals of *shraddh* and the ability of contact with the place to cause a *linga* to grow in the heart (Nautiyal 1994, 148–53). There are, in the geographically specific bulk of the text, numerous enjoinders for animal sacrifice, usually when the subject of the text is a form of Devi, the Goddess (Nautiyal 1994, 796, 749). The persistent presence of Bhairava and forms of the Goddess, particularly fierce forms such as Kali and Bhairavi, is a reminder of one of the common features of Himalayan religious cultures that marks Garhwal—the dense intermingling of Shaiva and Shakta traditions with the local deities of the region, deities who are often understood as forms of Bhairava or Devi. The Kedarnath locale, and *kedaramandala* more generally, are understood to be infused with tantric power. This is important to keep in mind because it helps in understanding why in and around Kedarnath the temple is not the sole focus of religious attention.

THE *KEDARAKALPA*

The final Kedarnath-related text to be considered is the *Kedarakalpa,* a late tantric text of uncertain date that exists in two published versions.[29] As discussed in the previous chapter, the *Kedarakalpa* serves as a guidebook for the reader whose goal is to walk into the physical Garhwal Himalaya and then, while remaining in his or her physical body, to walk to Kailasa and/or liberation. In this text it is possible to see how the efficacious power of the landscape and water accelerates the transformation of the devotee. In the Khemaraja Shri Krishnadasa edition of the *Kedarakalpa,* the text begins with four chapters of instructions for ritual preparation before commencing the main topic of the first half of the text: Shiva's description (first to Skanda and then to Parvati) of the immense potency of different sites in the Kedarnath area and the importance of drinking water (with numerous and diverse ritual specifications) at each site.

The story of the journey of a group of five tantric practitioners (sometimes called *saddhakas,* sometimes *siddhas*) who are walking from the world of death (Sanskrit: *mrtyuloka*) to the place where Shiva resides (Sanskrit: *Shivalaya*) is gradually woven into this overview of the potent ritual cartography of the Kedara region. The *saddhakas* drink water from important water sources and use the *aghora* mantra to protect them along the way in a journey that recalls that of the Pandavas. Jan Gonda (1963, 263) has translated the *aghora* mantra in this way: "Oṃ, be there adoration to thy reassuring manifestations, O Rudra, and to the terrific ones, to the (manifestations) which are (at the same time) reassuring and terrific, O Śarva, to all these (manifestations) in all respects." The *saddhakas* refuse the offers of several kings and deities to tarry in a specific realm, enjoy themselves for hundreds of years, and then be reborn into an advantageous birth. They keep their focus on Shiva and keep moving north. Thus the idea of walking to Kailasa in one's own body frames both the popular practice of Himalayan *yatra* and the more esoteric practice of journeying into the Himalaya armed with mantras. The emphasis through the text on the importance of drinking water in specific places is another embodied bridge between the esoteric and the exoteric that resonates through to the present day. This *yatra* into the *kedaramandala* can be understood as both inner journey and outer journey, both secret and public, for both the initiate and the noninitiate.

Returning to the *Kedarakalpa,* in the section of the text that specifically describes what one should do in Kedarnath proper we find the instruction that the visitor to Kedarnath should performatively become a bull and drink water, an act that recalls the practices of the Pashupatas discussed earlier in the chapter (*Kedarakalpa* 6.1–26, ed. Viśālmaṇi Śarmā Upādhyāy 1952, 24–29). After bathing in the Mandakini River and performing a series of ritual preparations, one is enjoined to face in the northeastern direction, drink water three times from the left hand and three times from the right hand, and then get down on all fours like a bull, make sounds like a bull, and "drink like a bull."[30] While doing this, one is to recite a Goddess-focused mantra that the Hindi translation is reluctant to reproduce, instead saying one should speak "unmanifest/unknown words" (Hindi: "*avyakt shabd kar ke*" in the Hindi commentary on lines 6.17–18) and then the words "I alone am Brahma, I am Vishnu, and thus also Shiva" (*Kedarakalpa* 6.19, ed. Viśālmaṇi Śarmā Upādhyāy 1952, 28). That is to say, the person ritually becomes Brahma, Vishnu, and Shiva while acting like a bull drinking water, an act with clear continuity to the earlier ritual practice frameworks of the Pashupatas.[31]

APPROACHING THE LORD OF KEDAR

Shiva's presence in Kedarnath is a self-manifest, pyramidal, mountain-shaped, rock-shaped place, a *linga* of light, and shaped like a buffalo. This presence does not lend itself to consistent description, but there is a persistent pattern, across time and language, to how different accounts of Kedarnath concentrate in a fuzzy and powerful way on the textures of how he is part of the place. We may conclude from the intricacy of rhetorical effort in this regard that as an abode of Shiva, Kedarnath poses and has posed a challenge to language. The significance and power of Kedarnath as a specific location spread out in variable fashion. In the *Kedarakhanda* numerous places of power in and around Kedarnath proper have the potential to become the objects of efficacious practice: rivers, *kunds*, valleys, *lingas*, rocks, and the environment more generally. The *Kedarakhanda* demonstrates a consistent refusal to distinguish in any final sense between Kedarnath and the region of Kedara (modern-day Garhwal) more generally, and it suggests a continuum of human, animal, and physical environment with significant overlap. This, I would argue, can be understood as a particular form of eco-social sentiment, a sentiment that regards human affairs as one part of a broader ecosystem. The *Kedarakhanda* invites the visitor to the region on a journey into a land seething with material religious power, some of which demonstrates personality and some of which does not. In the *Skandapurana* the presence of Shiva mingles with the waters of the Ganga and the efficacy of the Himalayan setting. The *Kedarakalpa* also emphasizes the power of flowing water in the Himalaya and reminds us that visitors to the region are not just traveling to see Shiva and be purified by the encounter but also in some sense on a journey to become the god himself.

Over the course of the accumulation of these historical and narrative layers we see several available models emerge for thinking through the form and flavor of human interactions with what is present at Kedarnath. Humans in Kedarnath may choose to think of themselves as following in the footsteps of the Pandavas, but the tone of the interactions of the Pandavas with Shiva in Kedarnath varies. Are they *dharmic* exemplars whose assiduous search for Shiva finally yields fruit, or are they seekers of liberation whose fierce determination and purpose lead them into an inappropriately intimate and violent encounter with Shiva that requires that they make redress? The *darshan* of Shiva in Kedarnath is the *darshan* of a deity whose face has turned away; the challenge posed by this unwillingness forces the Pandavas to resort to drastic measures, such as the threat of inappropriate behavior (forcing Shiva to pass between Bhima's legs, or striking him in frustration). Yet this Himalayan place of challenge enshrines a God who is hyperefficacious, hyperpresent, and able to cleanse *papa* that can be cleansed nowhere else.

Additionally, the Shaiva model allows the devotee to think of herself or himself as both creating a relationship with the god and attempting to become or join with the god, with all the complications that we now see this entails. Thus, when we imagine going to see Shiva in his Kedarnath abode, several questions arise. Are humans in Kedarnath on their way to becoming Shiva? How does the seeking of Shiva's *darshan* relate to somatically engaging the numerous powerfully charged sites in the Kedarnath valley? If the devotee wishes to become Shiva, and Shiva takes the form of a bull or a buffalo (as might the devotee as well, the *Kedarakalpa* reminds us), then what of the feeling of intense and sometimes violent (in the case of the blow from Bhima's mace) effort in many accounts of the Pandavas' interactions with Shiva when he is in his buffalo form? Are these not accounts that premise a devotional relationality of the human and the divine modeled in early form by the Pashupata, rather than the human becoming of Shiva suggested in the concept of *sarupata?* Is the wounding perhaps a self-wounding, a sacrifice of ego-based self? Does human presence in Kedarnath constitute progress toward greater intimacy and an eventual merging with Shiva in which inner and outer worlds collapse, as it might be for a Virashaiva devotee, or is it an opportunity to move beyond the relationality of devotion and to access liberating power inside of one's self by using tantric techniques grounded in a place replete with divine material power? Or is it something else entirely?

Attention to questions of form and presence, a disciplined focus on the materiality of the different variables present in the eco-social system of Kedarnath, starts to frame the underpinnings of the present in Kedarnath. We start to see how an ancient web of narrative, practice, real and imagined landscape, place, and Shiva came to be and how the enmeshing of Shiva with his Himalayan environment has influenced the web in ways that are dynamically contingent and that constitute the beginnings of a persistent pattern. We see how this web creates narrative possibilities for framing physical presence in the place that break down the boundaries between self, body, deity, place, and religious power. Entry into this world makes the experience of being human much more permeable and, relative to other forms of sentience and material embodiment, more fluid. We acquire a view of Garhwal in which divine powers are geographically ubiquitous—continuously, materially, and efficaciously emergent. We observe that Shiva-in-the-Himalaya and the Himalaya itself are simultaneously moral and not moral. They reward virtue and devotion, yet at the same time their inherent religious power is such that they affect the unprepared as well. Let us now follow these complex patterns and the power they produce into the next chapter.

3

Earlier Times

An excellent example of the power of the Himalayan place of Kedarnath as a whole is fictional, found in a short detective story written in Bengali by Satyajit Ray (2000) and translated into English by Gopa Majumdar: "Crime in Kedarnath." Ray wrote these stories as didactic travelogues about different regions in South Asia that educated younger readers about cultural diversity and history. While fictional, the stories skillfully and accurately portray the worldview and *mores* of his mid-twentieth-century, middle-class Bengali audience, who over the last century and a half have constituted one of the core visitor groups to Uttarakhand. In this episode the crime-solving trio of Feluda Mitter, Lalmohan Babu, and Feluda's nephew Tapesh are drawn to Kedarnath on a case and are speaking with a fellow Bengali they meet on the journey. He says to them:

> "I have been to Kedar and Badri twenty-three times. It's got nothing to do with religious devotion. I go back just to look at their natural beauty. If I didn't have a family, I'd quite happily live there. I have also been to Jamunotri, Gangotri, Gomukh, Panchakedar, and Vasukital. Allow me to introduce myself. I am Makhanlal Majumdar." Feluda said, "Namaskar" and introduced us. "Very pleased to meet you," said Mr. Majumdar. "A lot of people are going to all these places now, thanks to road transport. They are not pilgrims, they are picnickers. But, of course, buses and taxis can do nothing to spoil the glory of the Himalayas. The scenic beauty is absolutely incredible." (Ray 2000, 305)

Yet as the story continues it strikes a different tone:

> Ramwara [Rambara, the midway point between Gaurikund and Kedarnath] was at a height of 2500 meters. The scenery around us was absolutely fantastic. Lalmohan Babu went into raptures, recalling scenes from the *Mahabharata*. He declared

eventually that he would have no regret if he fell and died on the way, for no one could have a more glorious death. . . . In the remaining three and a half miles, only one thing happened that's worth mentioning. The tall spire of the temple of Kedarnath came suddenly into view after leaving Ramwara. Most of the travelers stopped, shouting, "Jai Kedar!" Some folded their hands and bowed, others lay prostrate on the ground. But only a few moments after we resumed walking it vanished behind a mountain. We could see it again only after reaching Kedarnath. I learned afterwards that the brief glimpse we had caught earlier was considered a special darshan. It was called *deo-dekhni*. . . . It was half past five in the evening by the time we reached Kedarnath. It had not yet started to get dark, and the mountain tops were all shining bright. It is impossible to describe what one feels on reaching a flat plateau after climbing uphill for several hours on a steep and narrow road. The feel uppermost in my mind was a mixture of disbelief, reassurance, and joy. With this came a sense of calm, peace, and humility. Perhaps it was those peaks which towered over everything else that made one feel so humble. Perhaps it was this feeling that evoked religious ardour, a reverence for the Creator.

A large number of people were sitting, standing, or lying on the rocky ground, overcome with emotion, unable to say or do anything except shout, "Jai Kedar!" . . . I had expected Lalmohan Babu to want to rest after our difficult journey. But he said he had never felt more invigorated in his life. "There is new life in every vein in my body," he said. "Tapesh, such is the magic of Kedar." (Ray 2000, 2:321–22)

The "magic of Kedar" described above results from the experience of the place as a whole combined with the difficult journey that precedes it. For the conversations undertaken in this book it may be viewed as an eco-social experience of Kedarnath that could, depending on the frame of analysis deployed, be understood as "religious" and/or Hindu. This example, at this moment in our historical narrative, is designed to provoke a question in the mind of the reader, a question that in colloquial Hindi is often expressed in this way: "Kahan se kahan tak?" (lit. "From where up to where?" or more figuratively "How on earth did we get from there to where we are now?"). By what processes did Kedarnath transform from a famous premodern place of pilgrimage in the Himalaya, inhabited by a powerful and complex god, to a site where a group of travelers might go together for a combination of reasons that include everything from religious devotion to a kind of adventurous curiosity that drives certain forms of leisure travel? In what follows I will trace how the complex premodern patterns charted in chapters 1 and 2 transformed into more recent patterns.

Religious travel to Kedarnath and to the Garhwal Himalaya more generally was part of the broader transformation of the somewhat specialized practice of *tirtha yatra* in premodern South Asia into a broad, popular practice of *yatra* supported by a growing rail infrastructure during the colonial period (Lochtefeld 2010). During this time and into the present, it has been well attested that pilgrimage and tourism have been increasingly conflated (Gladstone 2005). However, in Kedarnath and the Uttarakhand Char Dham, because of the relatively inaccessible Himalayan geography and the social history of Garhwal and Kumaon, these developments took a distinctive turn. Around the time of statehood in 2000, there was a massive rise in the number of *yatris* coming to Kedarnath, causing an accelerated spatialization of capital, a rapid influx of money connected to pilgrimage tourism that spread throughout the economic catchment area of the Uttarakhand Four Abode Pilgrimage to the shrines of Yamunotri, Gangotri, Kedarnath, and Badrinath.[1] This phenomenon ran precisely counter to the kind of sustainable Himalayan development imagined during the creation of the state in 2000. The earlier remoteness of the region intensified the impact of the changes.

Over time the Garhwali setting manifested a tense, emergent interplay of the local lifeworlds of small Himalayan valleys with larger-scale political, economic, ecological, and Puranic frameworks for thinking about the identity of the region. Arun Agrawal and K. Sivaramakrishnan (2003, 21) have characterized some aspects of what I am terming a "tense, emergent interplay" as a distinctively Uttarakhandi regional modernity in which the sense of modernity is "more tied to place, but the place-related ties themselves are produced by a belief in the political and economic discrimination faced by those living in Uttarakhand."[2] As part of this emergent "regional modernity," a regional understanding of the idea of "Himalayan nature" emerged, shaped in part by colonial and postcolonial discourses about the picturesque Himalaya, hotly contested practices of natural resource extraction, and the desire of hill peoples for social and political autonomy (Linkenbach 2006, 2007). Over time Kedarnath became one of the most prominent sites in the region to be associated with this set of discourses. The rising tide of pilgrimage tourism began to crest around the time of the creation of Uttaranchal/Uttarakhand as a separate state in 2000, and Kedarnath was one of the places where the interplay of these different forces was the tightest. Kedarnath valley locals began to lose touch with older ways of living and working in Kedarnath. In the Kedarnath valley, locals and visitors understood these changes as both arising from the power and importance of Kedarnath and at the same time detrimental to it. In these discourses, Kedarnath, whose power and importance were constituted by the enmeshing of Shiva with the natural Himalayan environment, pulsed as one of the centers of these webs of region and state. The diverse set of Shaiva and Shakta geographic imaginaires charted in the last chapter were partially decentered but remained influential and efficacious.

GARHWAL

From the thirteenth to the twentieth century, Garhwal differed from many other regions of the Indian subcontinent with regard to its position in political, cultural, and economic networks. Sites such as Kedarnath drew people from outside the region. On the basis of even a conservative historical assessment of the antiquity of the site, it is evident that from the thirteenth century onward Kedarnath was on the map of important *tirthas* visited by kings, renunciants, and those devotees with enough time and courage to make the journey. There are entries about *yatra* to Kedarnath in important *dharmashastra* digests (Vīramitrodayaḥ 1906; Lakṣmīdhara 1942; Kāṇe 1953). Beginning in the twelfth century, the rule of the Katyuri dynasty ended in Garhwal and Kumaon, and rulers from the Malla dynasty in western Nepal became active in the region. The area where Kedarnath is located underwent considerable political fragmentation. The next four centuries saw the practice of *yatra* to Kedarnath and Badrinath unfold primarily as a journey through a series of numerous small, competing principalities. Each of these principalities often centered on its own fort, or *gaṛh*, from whence comes the region's modern name of Garhwal, in Hindi, "the place of forts." This period ended with the consolidation of the Garhwal region into a single kingdom in approximately the sixteenth century by Ajay Pal of the Parmar/Panwar dynasty and the split of Garhwal and Kumaon into separate, competing regions (M. Joshi 1990, 63–64).

The relative geographic remoteness of the region may have been one of the factors that slowed the region's embrace of developments and networks originating from the plains. For example, it may have drawn out the processes through which Kedarnath became incorporated into transregional Shaiva and Puranic sacred networks. It also meant that traditional accounts of Kedarnath, in contradistinction to those of North Indian sites such as Braj or Somnath, do not include a narrative of destruction or the lapse of tradition sometimes associated with periods of Muslim rule. Rather, they often celebrate Shankaracarya's arrival in the region as rescuing the region from its control by Buddhists. Garhwal and Kumaon during these centuries became very involved in Indo-Tibetan commerce, and the transit fees from this trade and the pilgrimage activity of *yatris* were the major sources of income for the region (Rangan 2000). The Garhwal region also remained relatively politically autonomous during the Mughal period (Negi 2001, 17). The geography of the region generally and of the Kedarnath valley specifically meant that Garhwal (even compared to Kumaon) remained on the periphery, and out-of-the-way sites like Kedarnath even more so. While important for the *yatra* trade, the Kedarnath valley was relatively unimportant for commercial trade between Tibet, central Asia, and North India because the primary northern pass proceeds through Badrinath, whereas Kedarnath is closed off to the north by glaciers.

This geographic history has meant that in recent centuries Garhwal has exhibited a blend of local, animistic tradition with broader frameworks that today might be termed "Hindu and/or Buddhist," a mixture that is to widely varying degrees characteristic of a wide belt of Himalayan religious traditions and cultures stretching from modern-day Tajikstan, Afghanistan, and Pakistan to the northwest of India to China, Tibet, and Nepal in the north to Bhutan and the Indian state of Arunachal Pradesh in the east. Relative to the North Indian plains, the region known today as Uttarakhand has been a border zone where groups moving up from the North Indian plains mingled with arrivals from central Asia and indigenous groups already resident in the area. The worship of natural elements (Sanskrit: *bhuta*) in this region shades into shamanic interactions with place-specific local deities, which in turn shades into large-scale cosmological under-standings that include deities such as Vishnu, Shiva, and forms of Devi (Dhasmana 1995; Purohit, Negi, and Negi 1995). In Garhwal specifically, as opposed even to Kumaon and to many other regions in this Himalayan belt, there is additionally a distinctive depth to the overlap of local Himalayan traditions with the Sanskritic worlds of the *Veda*, the *Mahabharata-Ramayana,* and the *Puranas* (Sax 2011, 2002, 1991; Nautiyal 1994). This complicated social history also feeds into how different groups resident in what we now term Uttarakhand have historically viewed one another (M. Joshi 2011).

The region's relative isolation was breached when Gurkha kings of Nepal briefly and successfully annexed large portions of Garhwal and Kumaon in the late eighteenth century, beginning in Kumaon in 1790. The British were able to use the pretext of rescue from the occupation of Garhwal and Kumaon to develop a foothold in a region that contained an important route to Tibet, central Asia, and Russia through Badrinath. This meant that the region was both strategically important, because it connected northern India to Russia and China, and commercially important because it provided access to the Silk Road and contained the valuable natural resources of "*pashm* (Cashmere wool), gold, borax, and salt" (Rangan 2000, 73). After defeating the Gurkhas in 1815, the British annexed Kumaon and part of Garhwal, leaving what is now known as Tehri Garhwal as a princely state in the hands of the Garhwali king Sudarshan Shah, who subsequently established the capital of his kingdom in the town of Tehri.[3] The British struck a series of treaties with the Tehri court. Through these treaties, the British made sure that the *parganas* of Nagpur (containing Kedarnath) and Painkhanda (containing Badrinath) became part of British Garhwal rather than Tehri Garhwal and over the next century repeatedly rebuffed the Tehri court's requests that the administration of Kedarnath and Badrinath be fully in royal control (India Office 1895; Traill 1823; Crown Representative's Records 1936; India Office 1942, 46; Rawat 2002, 19–20; Negi 2001, 24–25). British Garhwal itself became part of the province of Kumaon. Garhwal's geographic autonomy from Mughal rule notwithstanding, Garhwali political administration incorporated standard Mughal administrative patterns such as the *pargana-patti* system of land division and taxation that served as the basis for this division (Gordon 1985). This became the basis for tax collection in British Garhwal, and the Kedarnath valley came under British rule.

As rulers who were concerned with both profit and the preservation of their version of order, the British had several areas of particular interest in British

Garhwal. Haripriya Rangan (2000, 70–71) has observed that in the "regional economy of Garhwal" in the eighteenth century "transit duties" generated from *yatris* were one of the most significant forms of revenue in the region, along with export and transit taxes charged for mining and the extraction of other resources like timber. The British aimed to build on this commercial characteristic of the region. This focus on resource extraction took several early forms: an emphasis on the production of tea, hemp, sugarcane, cotton, and rice. In the second half of the nineteenth century, after the formal declaration of direct rule by the British government in 1858, Garhwal was no longer as profitable a site for these agricultural products. The emphasis in British Garhwal shifted to what had become the only commercially viable natural resource that could be profitably extracted from Garhwal at the time: timber. These large-scale extractive relationships to *kedaramandala* mixed in with the older, diverse Shaiva, Shakta, and Himalayan ways of conceptualizing, experiencing, and enacting connection with Himalayan landscapes, and resistance to these extractive movements drew to some degree on these older, already emplaced patterns.

The impact of British colonial presence on the natural environment was already predominant in a story that, in the nineteenth, twentieth, and now twenty-first centuries, has become depressingly familiar—a story of resource extraction, widespread deforestation, and damming of rivers, and in a more general sense the large-scale transformation of human relationships to the natural world into relationships characterized by a sense of often short-term instrumentality. These relationships in turn refract through the pernicious prisms of large-scale industrialization and the discourses and practice of development, something I address more substantively in chapter 6. In a region like Garhwal this has meant that over a period of two hundred years a pattern built up in which Garhwalis became the often unwilling providers of resources for people living outside Garhwal in ways that were profoundly disadvantageous.

FORESTS

The preservation, management, and sale of timber would be of central concern to the British and later Indian governments for the next century and a half. The new Indian rail network (begun in 1864) was one of the catalysts behind the development of a Forestry Department whose primary aim was to manage, preserve, and extract timber to be used in the construction of railway sleeper cars (Guha 2000, 37). The creation of a Forestry Department set in motion an almost century-long conflict. On one side were timber companies from outside the region and the colonial and postcolonial iterations of local, regional, and national governments who supported them along with specific organs of the government such as the Forestry Department. On the other side were local

Garhwalis and Kumaonis, notably the local women who were the primary gatherers of wood. They resented how the Forestry Department took over what had been commonly held land, cut down far too many trees, encouraged the growth of the kinds of trees that would support the timber industry, and denied locals access to what had formerly been their own local natural resources. Extralocal control and cutting of trees also challenged the political and religious authority of local deities who "owned" particular territories in which forests were located (A. Kumar 2011, 90.[4] These tensions, along other causes for resentment against British rule, occasionally crystallized into demonstrations, strikes, and protests, beginning in the late nineteenth century (Rawat 2002, 130–51). After Independence the Indian Forestry Department continued down the same path (Gadgil and Guha 1995, 23). These tensions eventually created the conditions for what would emerge in the 1970s as the Chipko movement, now world-famous for their very successful use of the nonviolent, Gandhi-inspired technique of tree hugging (the name Chipko derives from the Hindi verb *chipakna*: cling, stick, adhere) as a way to protest the management and felling of local trees by the Forestry Department and commercial timber companies. This movement was championed by environmentalists such as Chandi Prasad Bhatt, Sunderlal Bahuguna, and Vandana Shiva and was famously spearheaded by local women.[5] The village of Fata in the Kedarnath valley, just north of the temple of Mahismardini, was one of the Chipko movement's important early sites of nonviolent resistance (Guha 2000, 157–58). Today Fata is one of the places in the Kedarnath valley where one can board a helicopter to fly to Kedarnath.

RIVERS

A second focus of the British in the region was the control of rivers. In the mid-nineteenth century the British decision to canalize the Ganges River near Haridwar to alleviate the impact of a recent large-scale drought on the Indo-Gangetic plain led to wide-scale protests in Haridwar and caused many local groups to unite against this British initiative (Lochtefeld 2010, 88–96; Rangan 2000, 87–90). Approximately a century later, the decision of the Uttar Pradesh and central governments to begin construction of a large-scale hydroelectric dam on what becomes the Ganges River just below Uttarkashi in Tehri Garhwal in order to generate electricity and provide water for use outside the region raised another round of massive, ongoing protest and controversy. The Tehri Dam is a massive civil engineering project, a structure of staggering scale located in a region well known for its historic and potential future seismic activity. Its construction, symbolic of a Nehruvian vision of large-scale development, was seen to be of little benefit to Garhwalis themselves. Emma Mawdsley (2005, 9) observed that the "displacement from land" and "environmental damage" created by the construction of the dam disproportionately affected predictably vulnerable groups: "the poor, women and rural people of this marginal mountain region." Resentment against hydroelectric projects on the Ganga continued to build and diversify. One of the most recent chapters in this story, chronicled by Georgina Drew (2017, 14), has been the surge of environmental activism that produced the central Indian government's designation of part of the Bhagirathi Ganga as an "Ecologically Sensitive Zone."

Haripriya Rangan (2004, 213–24) observes that a lack of sufficient government relief for the destruction caused by serious floods, landslides, and erosion (particularly in the 1970s), characteristic of larger trends in the "economic marginalization" of people in the Garhwal Himalaya, was a tragic consequence of this national development model that fueled considerable regional resentment. Anger against such large-scale hydroelectric projects overlaid and joined the nexus of resentment, social and economic contestation, gendered resistance, and allegiance to local and regional patterns of Himalayan religious sentiment connected to the battle over timber. Linked concern about control of, and respect for, trees and flowing water and the small-scale and sustainable ways of being associated with such respect influenced the vision for the separate state of Uttarakhand, but the continuation of destructive and environmentally un-sustainable practices in the new state undermined that vision and fueled a sense of betrayal brought to crisis point by the floods of 2013 and the response of the state government in the aftermath.

These colonial and postcolonial developments joined the Himalayan landscape conversation (recalling Ingold's idea of "landscape" as an emergent conversation between human and other types of interdependent organisms all based in the natural world), a conversation already infused with diverse religious understandings of the enmeshing of Shiva with the Himalayan terrain and his close relationship with the flowing waters of the Goddess and the pervasive presence of local deities. The *shakti* of Shiva, of the rivers, and of the Himalaya, distinctively resident as it was in Kedarnath, began to be treated as a natural resource to be mined and controlled and pictured. But another important change was building force as well: the emergence of a regionally specific set of discourses about the idea of "nature" (Hindi: *prakriti*) that further framed the production of landscape, identity, and the experience of place in Garhwal, especially in places such as Kedarnath. I return to these developments later in the book in the context of a discussion that looks at how understandings of "nature" overlap with the divine powers of Shiva and Devi.

YATRA REGULATION AND TRANSPORTATION

The British government in India was deeply involved with the regulation and support of pilgrimage across the subcontinent in numerous ways, an endeavor closely bound up with the control and management of fields, forests, and rivers because of the necessity for transport infrastructure in the form of roads and railways and the tax revenue to fund that infrastructure. The Archaeological Survey of India and its practice of what Toni Huber (2008, 252–53) called the "distinctive colonial phenomenon of monumental archaeology" was, for example, a primary catalyst for the renewed importance of Bodh Gaya as a site of Buddhist pilgrimage for Buddhists outside the Indian subcontinent. The assembling of Indian Muslims for travel on the Hajj pilgrimage to Mecca, and the potential of the transmission of cholera from Europe to India via Hajj travel to and from Mecca, produced legislation aimed at controlling these politically suspicious potential problems (Mishra 2011, 15–19). Disease prevention was also of concern for the government of British Garhwal, as was the regulation of sanitation at pilgrimage places more generally (Khalid 2008). Assembly at large *melas* such as the Kumbha *mela* was viewed with suspicion for the same reasons (Mishra 2011, 15; Maclean 2008, 12). Groups of Shaiva ascetics were of British governmental concern at this time because they were seen as sources of conflict, axes of political power, and a potential pool of military recruits (Pinch 2006, 77).

The increased number of *yatris* in the nineteenth century was one of the indications that *yatra* in South Asia had begun to change from an elite and renunciant practice to something that regularly included members of society from diverse social and economic groups. This change, which in some cases might have meant a threefold increase in the number of visitors to specific places of religious significance, was facilitated by better transportation, the abolition of "pilgrim taxes," the rise of a new kind of "conspicuous piety" by kings, and "new men who built up their fortunes through the service of the British" (Yang 1998, 134; Bayly 1988, 159). Significantly, building on an observation by Ashish Nandy (1980, 7) about nineteenth-century views of the contested practice of *sati*, Yang (1998, 138–39) observes that the act of pilgrimage "may even have gained in status and currency over the course of the colonial era" and "may well have represented a way of expressing 'conformity to older norms at a time when these norms had become shaky within'" (see also see also Maclean 2003, 891). Yang (1998, 137) observes that the rise in the popularity of pilgrimage centers in the nineteenth century also produced new forms of religious advertisement: vernacular renderings of Sanskrit works celebrating the greatness of specific deities and deity-related places (Sanskrit: *mahatmya*) in the form of printed pamphlets.

THE RELIGIOUS ENDOWMENTS ACT

The money associated with important pilgrimage places created many entanglements for British administrators and for the British government more generally. The massive amounts of revenue from land owned by temple deities and from funds and goods donated to temples produced numerous situations where colonial administrators found themselves, first in an ad hoc modality and then in an increasingly formal way, embedded in Hindu temple administration in order to realize profits from revenue collection and to adjudicate disputes between local stakeholders for whom they had become another group of local and regionally powerful political actors (Yang 1998, 134; Presler 1987, 15; Appadurai 1981). In these contexts the British found themselves struggling between conflicting policy imperatives: the responsibility (once that of the king) to maintain order and to ensure the proper handling of temple funds and the desire not to interfere in religious matters (Presler 1987, 15–56; Appadurai 1981, 105–38; Dube 2001; Reddy 2014, 147–95).

From 1817 to 1839 the East India Company formally involved itself in the financial and administrative management of temples and mosques (Mudaliar 1976, 22; Presler 1987, 16–17). Then in 1839 it attempted to extricate itself from this involvement and practice a "policy of neutrality towards religious institutions" (Mudaliar 1976, 22; Presler 1987, 19). In the middle of the nineteenth century, when Britain formally annexed India as part of the British Empire, this situation changed again. The British government attempted to officially and legally withdraw from formal involvement with Hindu temple administration and instead to place temple administration in the hands of locally constituted committees or "area committees" of temple trustees drawn from among the different groups of local stakeholders with traditional claims to administrative authority. The operating procedures of this new organizational scheme were meant to ensure the proper management of temple funds generated by property revenue, donations, and fees given for ritual services. The formal legal mechanism of this withdrawal was the Religious Endowments Act XX of 1863 (Mudaliar 1976, 22–23). However, this act did not immediately obtain in all cases. Rather, the state would create an amendment that would extend and shape the act so that it could apply, with stipulations that made sense for the specific case, to a particular temple or set of temples. Initially this happened when a complaint regarding temple management rose to the level of a civil suit against temple trustees or members of the local "area committees" (Presler 1987, 24–25). As a result of this tortured dance of power, policy, law, and money, there were numerous multiact legal dramas involving kings, priests, and colonial administrators over the course of approximately two centuries. In the case of Badrinath and Kedarnath, this process began with the Shri Badrinath and Kedarnath Temples Act of 1939 (Husain 1965; Badri-Kedar Temple Committee n.d.).[6] According to the "prefatory note" of the act, which was later extended to include Kedarnath,

> The Badrinath Temple which is one of the foremost sacred places of Hindu pilgrimage in India is situated in the Garhwal district on the heights of the Himalaya. Under the scheme of 1899 at present in force its management is in the hands of the Rawal, while the Tehri Durbar is invested with certain supervisory powers. The defective

nature of the scheme has been the source of constant friction between the Rawal and the Tehri Durbar. As a result, supervision of the temple has suffered, its income has been squandered and the convenience of the pilgrims has been neglected. The unsatisfactory condition of the temple which has existed for a long time was specially brought to the notice of Government by the Hindu Religious and Charitable Endowments Committee in 1928. Since then public agitation has been continually pressing for reform in its management. The Bill which is now introduced seeks to remove the chief defects of the present scheme. It restricts the Rawal to his priestly duties and places the secular management of the temple in the hands of a small Committee which would be partly elected and partly nominated. It preserves at the same time the traditional control of the Tehri Durbar; while adequate powers have been reserved for Government to guard against mismanagement by the Committee.

Later on, the act clearly defines "Temple" as referring to both the "Temple of Shri Badrinath and the Temple of Shri Kedarnath" (Husain 1965, 525), Today the Samiti generates its own revenue and has a board composed of selected and elected officials, some of whom must be from particular regions of Garhwal. It runs the Badrinath and Kedarnath temples, along with a host of ancillary sites. It also funds and manages several educational charitable institutions, such as Sanskrit and Ayurvedic colleges in the Kedarnath valley. The disposition of the rights and authority of the *tirth purohits* in this document and its later amendments led to over a century of contention between the Samiti, the community of *tirth purohits,* and to some degree the *rawal* over the rights of *tirth purohits* to collect the ritual fees (Hindi: *dan* and *dakshina*) associated with the performance of *pujas* and other ritual services in and around the temple. The celebration of a successful appeal by the *tirth purohit* community in this regard was one of the important events of the 2007 season that I spent in Kedarnath (Tandon 2007; India Office 1942; Mukherjea, Ali, and Das 1952).

THE GROWTH AND CHANGE OF *YATRA* IN GARHWAL

The general trends in temple administration and the popularization of *yatra* practice apply to the case of Garhwal. James Lochtefeld (2010, 70–71, 255) has observed that it was the completion of the Upper Ganges Canal in 1854, followed by the inclusion of Haridwar on the Oudh and Rohilkhand Railway in 1886, that would transform Haridwar into a pilgrimage place of year-round importance. This development also made it much easier to begin a *yatra* to Kedarnath and Badrinath on roads maintained by British administrators. In 2007 and 2008 Rishikesh, just northeast of Haridwar, was the most common point of departure for most groups undertaking the Uttarakhand Char Dham Yatra. British administrators expanded and systematized two related systems of land taxation connected to Kedarnath and Badrinath: *gunth* and *sadavart/sadabart. Gunth* refers to lands whose product belongs to the temple as the result of a donation, usually from a king. *Sadavart* "is the term applied to an endowment provided by the land revenue of assigned villages, originally for the purposes of providing with food indigent pilgrims visiting the shrines of Kedarnath and Badrinath" (Walton [1910] 1989, 104). The British used these funds to build and improve roads and provide medical dispensaries along the pilgrimage routes to Kedarnath and Badrinath as a bulwark against the transmission of epidemics. It was during the nineteenth century that *yatra* to the Uttarakhand Himalaya began to occupy the double space of mass pilgrimage and perilous undertaking from which it now continues to emerge.

George William Traill (1823, 119–20), a British official who worked as a commissioner in parts of Garhwal and Kumaon, estimated that Badrinath could receive up to ten thousand visitors in a normal year and up to twenty thousand in a year such as 1820, when the Kumbha Mela or Ardha Kumbha Mela was taking place, with the arrival of many more prevented only by a cholera epidemic. The adventurer T. J. Saunders (1844, 64) claimed that approximately fifteen to twenty thousand *yatris* visited Kedarnath annually. His assessment about the impact of the recent road (path) built for *yatris* to Kedarnath by George Traill was prescient, deeply troubling, and ultimately wrong:

> The new road has now rendered the temples accessible to all, and in time this very facility of reaching them, which at first you would be disposed to say would be the instrument of increasing and extending fanaticism and idolatry, will, to a moral certainty, tend more than anything else to their overthrow. Juggernath when idolatry was taxed and difficult of approach, was far more popular than it is now; and in half a century, when Kedarnath and Budreenath become better known to the multitude, the pilgrimages for martyrdom by cold and privation will gradually diminish in number, and be succeeded by those of enthusiastic travellers, like ourselves, who undertake this journey of endless toil merely to have an opportunity of admiring the stupendous grandeur of the regions of eternal winter. Mr. Traill, by his removal of the great obstacle in the way of a safe pilgrimage to Kedarnath and Budreenath, hazarded his reputation as a Christian, and subjected himself to the imputation of being an encourager of pagan idolatry.

Saunders claimed that such criticisms misunderstood what he took to be Traill's ultimate purpose: "The surest way of letting idolatry die a natural death, is to make it cheap, and common, and easy of access" (64).

Better transport and the democratization of pilgrimage as a form of public piety saw more visitors of increasingly diverse backgrounds traveling to Kedarnath. The gazetteer writer Walton ([1910] 1989, 56–57), quoting and adding to the account of Edward Atkinson (1886), wrote about Kedarnath:

> The pilgrims number 50,000 or 60,000 yearly and come from all parts of India. Formerly devotees used to immolate themselves from the Bhairab Jhap near the temple of Kedarnath, and to the present day an occasional enthusiast wanders blindly up the eternal snows seeking the heaven of the gods. "A popular belief exists that Siva

frequently makes himself visible on the crest of the great peak and that the wreaths of smoke seen there from below are not the result of whirlwinds gathering up the finer particles of snow, but the smoke of sacrifice made by some highly favoured follower."

Many visitors understood these high regions to be the assembly place of gods, and the natural sounds of trees and avalanches to be the sounds of their activities. The full sensory impact of the region was olfactory as well. Quoting Atkinson, Walton ([1910] 1989, 57) relates that "the sweet smelling flowers and other vegetation found near the limits of eternal snow frequently overpower the traveler and combined with the rarefaction of the air cause a faintness which may be attributed to superhuman powers." These remarks underscore how the physical environment was understood to signify materially and sensuously the presence of Shiva and other divine powers in the region beyond the temple and in the landscape itself. They also recall the life-ending journey of the Pandavas and highlight the characteristic overemphasis that colonial discourses often gave to forms of religious practice they found to be culturally challenging (Dube 2001, 128–29). Walton ([1910] 1989, 176) also briefly describes the practices of *yatris* in Kedarnath: "The ceremonies to be observed by the pilgrims are very simple, consisting of a few prostrations, an embrace of the linga and the hearing of a short ritual and discourse from the officiating priest. The pilgrim carries away in copper jars from the sacred pool some water which is highly charged with iron and Sulphur."[7] Also worthy of note is the account of the visit of Sister Nivedita, born Margaret Noble, "the young Irish disciple of the nineteenth-century Hindu reformist and proto-nationalist Swami Vivekananda" (Roy 2006, 498). Nivedita's account (published in 1928), making up approximately half of an eighty-six-page account of her journey to Kedarnath and Badrinath, blends her wonder at the natural environment, her Orientalist and religious wonder at the ancient richness of Indic pilgrimage traditions, her early nationalist sentiments, and a scholarly interest in the history of Indian religions. She consistently makes assertions regarding the pre- and post-Shankarite nature of the Shaiva sites she sees on her journey and makes numerous observations regarding the traces of Buddhist presence in Garhwal left in the archaeological record of the temples she visits.[8] She marvels at the number of *yatris*, and especially at how many of them are women. She writes, "Who uttered a doubt that India had a place and a life for women? Certainly none who had ever seen a pilgrimage" (Nivedita 1928, 4). In her vibrant and lengthy description of her actual time in Kedarnath, she offers a substantial description of Kedarnath *darshan* and *arati*:

> Suddenly we were called to see the *arati*. Darkness had fallen but the mists were gone, and the stars and the snows were clear and bright. Lights were blazing and bells clanging within the temple and we stood without, amongst the watching people. As the lights ceased to swing and the *arati* ended, a shout of rapture went up from the waiting crowd. Then the cry went out to clear the road, and the rush of the pilgrims

up the steep steps began. What a sight was this! On and on, up and up, they came, crowding, breathless, almost struggling, in their mad anxiety to enter the shrine, reach the image, and at last, by way or worship, to bend forward and touch with the heart, the sacred point of the mountain! For this half-embrace is what the worship consists of at Kedar Nath . . . It was one of the sights of a life time, or [sic] to stand there, and watch the pilgrims streaming in. It seemed as if all India lay stretched before One, and Kedar Nath were its apex, while from all parts everywhere, by every road, one could see the people streaming onward, battling forward climbing their way up all for what?—for nothing else than to touch God! (40–41)

The peripatetic polymath Rahul Sankrtyayan, father of the Hindi travelogue genre, dwelt on another aspect of *yatra* to Kedarnath in the early to mid-twentieth century: the difficulty of the ascent. He remarked that, before the improvement to the path made by the British, it was easy to imagine that some *yatris* visited Kedarnath visited with the intention of "ascending to heaven" in the manner of the Pandavas—that is to say ending their life in or behind Kedarnath—because the combination of altitude and weather made the ascent so difficult and death seem so close (Sāṅkṛtyāyana 1953, 424–25). Sankrtyayan's pithy observation, which drew on the anecdotally attested earlier twentieth-century phenomenon in which some *yatris* would celebrate their own funeral ceremonies before setting out on *yatra* to the central Indian Himalaya because they did not think they would return (424), neatly summarizes the ways that preexisting practices, ideas, and stories connected to Kedarnath were changing form.

HILL POLITICS

These *yatra*-related changes largely involved the way the region was changing to accommodate visitors and the number and diversity of visitors arriving in the region. Corresponding political regional changes are also part of this story. Parallel to the trajectories charted in this chapter thus far were political processes that, after Independence, placed the formerly competitive mountain regions of Garhwal and Kumaon in solidarity with each other as the two hill areas within the large (and mostly plains-based) state of Uttar Pradesh. While there had been demand from "feudal elites" well prior to Indian independence in 1947 for the creation of separate mountain provinces (for example, the demand for a Kumaon Province in 1916) and from "urban elites" in the 1950s, 1960s, and 1970s, these attempts were marked by a lack of grassroots support and competition between Garhwal and Kumaon (A. Kumar 2011, 88–89). As Anup Kumar (2011, 102) notes, this began to change when hill people began to join together in "socio-ecological movements such as the . . . University Movement (1971–1973), the Chipko Movement (1973–1980), Nasha Nahi Rozgar Do (Give Jobs Not Intoxication, 1983–1985), Anti-Tehri Dam and the Askot-Arakot Abhiyaan (Askot-Arakot Foot March)."

These movements, rooted as they were in grassroots concerns about the relationship of local living conditions to the needs and desires of people living outside the region, laid the groundwork for what would become a *jan andolan,* a people's movement for the creation of a separate state. The connections of residents in the region to the trees and rivers of their own lifeworlds, combined with many different forms of economic, social, and political discrimination they endured at the hands of representatives of the state government, gave these movements a new, unifying power. There was an explosion of activist groups composed of students and/or women with links to other activist groups around the new nation. As these movements gathered steam and as the political ecology of the region continued to fuel the concerns of those involved, the drive for regional autonomy began to be in increased tension with the political vision of the national Congress Party, which had been in power in Uttar Pradesh for decades. A new political party emerged in 1979, the Uttarakhand Revolution Party (Hindi: Uttarakhand Kranti Dal), which allied itself with the national parties who were in competition with the Congress Party: first the Janata Dal and then the Bharatiya Janata Party (BJP) in 1989 (A. Kumar 2011, 116–17). In 1989, in a reflection of changing national political trends, a Janata Dal and BJP-led coalition came into power in Uttar Pradesh and made regionalism part of their political platform, including the passage of an Uttaranchal Statehood Bill in 1991 that recommended a statehood separate from Uttar Pradesh; this was the year that a destructive earthquake in the Uttarkashi district of Garhwal made many *paharis* feel that their plains-based government did not have the requisite knowledge and commitment (118). Kumar relates, in response to a serious earthquake that struck central Garhwal in 1991, that "the earthquake killed about a thousand people and rendered many more homeless. The official response to the Uttarkashi earthquake became the exhibit one of the aloofness of the government in Lucknow." He further observes that "in public discussions the earthquake brought the people and their socio-ecological concerns together in a moment of crisis" (121–22). These regionally focused changes took place in the context of a national conversation shaped by the Mandal Commission about the creation of legal protections for underprivileged castes and tribes and a rising tide of Hindu nationalism that crested with the destruction of the Babri mosque in Ayodhya, Uttar Pradesh, in 1992.

PRECARITY AND SOLIDARITY

It was not incidental that governmental response to an earthquake was an important political catalyst to regional "socio-ecological" solidarity. Phenomena such as earthquakes, landslides, and floods have been part of the cultural and physical terrain in the Garhwal Himalaya for a long time. In addition to the severe earthquake in Uttarkashi in 1999, in 1991 Chamoli was at the epicenter of a large and destructive quake. The entire region was shaken by an earthquake in 1803, with Uttarkashi, then known by its older name of Barahat, again experiencing terrible damage. It is reasonable to estimate that similar earthquakes have been occurring periodically in the Himalaya for much longer than these recently documented events. Roger Bilham (2004, 842) begins the story of attested earthquakes in the Himalaya with what he terms "a probable Himalayan earthquake reputed to have occurred during the time of enlightenment of Buddha *ca.* 538 B.C." More locally, in 2014 there was a landslide in the Madmaheswar valley near Ukhimath in which debris flow decimated almost the entire village of Mongoli, a village already affected by the Kedarnath floods in 2013. Major landslides are an annual occurrence whose incidence and intensity rise when there is an earthquake or flood event; in recent decades, they have become relatively more common because of ill-planned cutting and widening of roads (Barnard et al. 2001). Intense rainfalls have been triggering floods and landslides for centuries (Wasson et al. 2013). In the last century, however, the incidence of Glacial Lake Outburst Floods (GLOFs) has been on the rise as increased glacial melt has begun to produce more glacial lakes (Richardson and Reynolds 2000). Small-scale landslides are so common as to be simply part of everyday routine. Bus and jeep passengers in Garhwal have a system for landslide blockage—the passengers on each side of the blocked road will simply walk across the blocked area and get on a jeep or bus that, once it has turned around, will be able to proceed. A friend of mine once described taking a bus ride in Garhwal as a "game of destiny" (Hindi: *kismat ka khel*).

The Garhwal Himalaya is sometimes described as "fragile," in part because of all of this geological activity (Rizvi 1981).[9] But *fragile* is something of a human-centric misnomer that misrepresents both the problems that humans face when living in changeable terrain and the fact that the actions of humans can clearly change that changeable terrain, even in the short term. Change is not necessarily fragility. Historically, humans in Garhwal who have had to live in intimate relation to these conditions developed strategies for successful living (such as not building houses on landslide-prone mountainsides) and, on a range of levels from the economic to the existential, were forced to take into account the inevitability of floods, landslides, and occasionally earthquakes. Vandana Shiva (1988, 184) reads the myth of the descent of the Ganga as a metaphorical expression of this situation of necessary sensitivity to the natural environment, seeing in the story a "description of the hydrological problems associated with the descent of mighty rivers like the Ganga, which are fed by seasonal and powerful monsoonic rains" whose forces are mitigated by the mountain forests that are Shiva's hair. A deep-rooted sense of contingency comes with living in this Himalayan territory, or in any situation that is marked in the long term by what Jessica Lehman (2014) and many other geographers term "uncertainty." *Paharis* who have no choice but to deal with this set of conditions experience a sense of solidarity with other *paharis*.

The response of the Uttar Pradesh government to the Uttarkashi earthquake of 1991 did not demonstrate a sense of empathy or knowledge about what this event felt like for *paharis,* so it contributed to currents of socioecological solidarity and resentment against non-*pahari* outsiders that were already running high.

A NEW STATE

In 1994 the state government announced new quotas for the admission of historically disadvantaged groups (Other Backward Classes) to state schools and universities. Many, notably students, felt that the new quotas did not fairly reflect the actual percentage of Other Backward Classes present in the mountainous parts of the state. Hunger strikes led to arrests, and arrests led to a shutdown of schools. Students in Srinagar blockaded traffic on the main pilgrimage road, and women joined the students, connecting the protests to the long-standing tradition of female activism in the region (A. Kumar 2011, 135–37). Protests, hunger strikes, and police reprisals spread throughout the region and began to center in Dehra Dun. Protesters were shot in Mussoorie (Garhwal) and Khatima (Kumaon). Eventually one organization, the Uttarakhand Progressive Women's Forum, decided to stage a rally in New Delhi. Activists from all over the region began to converge on Delhi but met with resistance and blockades by the police, most famously at Rampur-ka-Tiraha on the outskirts of Muzzafarnagar city, where an escalation of tension between the protesters and the police became violent, resulting in the death of "about fifteen protesters" and "hundreds" of injuries (A. Kumar 2011, 205). When national newspapers reported that "policemen had assaulted and allegedly raped some women" during the confrontation, riots broke out and "The government imposed 24-hour curfew all over the Uttarakhand region" (A. Kumar 2011, 214). These events created new political momentum for regional autonomy.

In 1996 the BJP became the dominant party in the "mountain districts" in Uttar Pradesh, and in 1997 a BJP-led coalition government called the National Democratic Alliance came to power in the state on a platform that included the creation of Uttaranchal (which would later become Uttarakhand). Uttarakhand and the other two states undergoing similar processes (Jharkand and Chhattisgarh) successfully became states in 2000. The founding vision of Uttarakhand (Uttaranchal) was in crucial ways premised on the commitment to greater sensitivity toward the local living conditions of hill peoples (Hindi: *pahari*), conditions that are often framed using the idea of "remoteness" (Mathur 2015b). This happened at roughly the time that the region had begun to see an increase in the number of middle-class *yatris,* the result of an array of developments correlating in important ways to the rise of the middle class in India in the 1980s and then after economic liberalization in the early 1990s (Sridharan 2004; Brosius 2010).

NOSTALGIA FOR EARLIER TIMES

When I discussed twentieth-century history in Kedarnath, people typically did not volunteer opinions, memories, or analyses about the history of the new state and the vision on which it was based. In retrospect, I wonder what I would have heard had I pressed the issue. But political history was not the story that seemed to matter in my conversations about recent history in Kedarnath. What mattered was how Kedarnath itself had been "in earlier times" (Hindi: *pehle zamanon men*), hence the title of this chapter. During my fieldwork in 2007 and 2008, older *tirth purohits* would often begin interviews and conversations by telling me their memories of earlier times and their parents' and grandparents' stories about how Kedarnath used to be. When the village was less developed (even only decades ago, they would say), Kedarnath at night was in many respects a fearful place, with most locals unwilling to leave the far smaller enclosure of the built environment of the village during the night for fear that they would attract malevolent supernatural attention. Elderly *tirth purohits* recall a time when *yatris* would usually stop for a night halt in Rambara and then go up to Kedarnath and return to Rambara in the same day without even spending the night. The relationship of *tirth purohit* to patron (Hindi: *yajman*) in the Kedarnath of these earlier times was, according to these nostalgic memories, a far more intimate and personalized relationship in which the *yatri/yajman* depended on the *tirth purohit* for guidance, hospitality, education about Kedarnath, the performance of all ritual needs, and food. The *tirth purohit*, in accordance with his traditional identity and rights (Hindi: *adhikar*), was happy to provide these necessary services in unstinting fashion. This was contrasted with the environment of recent years, when the massive increase in the number of *yatris* was beginning to make this kind of traditional relationship optional and people were not afraid to walk beyond the borders of the Kedarnath village at night. Looking back on conversations about changes in Kedarnath, I have realized that, while the political history of the new state was not explicitly mentioned very often or explicitly correlated to these changes, the sense of nostalgia that came through in these conversations was distinctively regional, a *pahari* nostalgia.

What were the contexts for these nostalgic memories? The development of motor roads in the Kedarnath valley received special attention as a result of the Sino-Indian border conflict in the 1960s, and, just as it had done before, easier access brought more visitors. By approximately the beginning of the sixties, the motor road had reached Guptkashi (Ḍabarāl, n.d., 241). The historian Shivprasad Dabral estimated that approximately one hundred thousand (one *lakh*) people a year were going to Badrinath just after the middle of the twentieth century (576). In recent decades the number of visitors to Badrinath has oscillated between approximately two and three times the number of visitors to Kedarnath. It is a safe assumption that Kedarnath also received fewer visitors in earlier times, thus giving

TABLE 2. Annual number of visitors to Badrinath and Kedarnath, 1987–2018

Year	Badrinath	Kedarnath
1987	271,850	87,629
1988	372,772	137,095
1989	370,820	115,081
1990	362,757	117,774
1991	355,772	118,750
1992	412,597	141,704
1993	476,523	118,659
1994	347,415	104,639
1995	461,435	105,160
1996	465,992	105,693
1997	361,313	60,500
1998	340,510	82,000
1999	340,100	80,090
2000	735,200	215,270
2001	422,647	119,980
2002	448,517	169,217
2003	580,913	280,243
2004	493,914	274,489
2005	566,524	390,156
2006	741,256	485,464
2007	901,262	557,923
2008	911,262	470,048
2009	916,925	403,636
2010	921,950	400,014
2011	980,667	571,583
2012	985,631	583,176
2013	497,744	312,201
2014	180,000	40,832
2015	359,146	154,430
2016	624,745	309,746
2017	884,788	471,235
2018 (till May 3)	71,739	62,535

SOURCE: Badri-Kedar Temple Committee (n.d.).

us an idea of how to apply these numbers. The increased ease of access yielded by the extension of drivable roads and attendant infrastructure saw the continued increase in the numbers of *yatris* journeying to Kedarnath each year (table 2).

As the table shows, the number of visitors to Kedarnath took a fairly sharp jump in 2000, the year that Uttarakhand (first as Uttaranchal) became its own state, separate from Uttar Pradesh. Vishwambhar Prasad Sati (2013, 101) observed a 25 percent increase in the number of visitors to Kedarnath between 2000 and 2010. In 2014, by comparison, the state limited the number of *yatris* per day to five hundred, but the volume of visitor traffic never made enforcing this regulation a problem. The drop in 2013 and 2014 tells its own terrible story. The resurgence that began in 2015 testifies to a humbling combination of human will, economic incentive, and faith.

YATRA TOURISM

In the mountainous terrain that constituted the majority of the new state, pilgrimage tourism became one of the most important and optimistically regarded factors in the new state-based regional economy (Mamgain 2004, 256–57; 2008). In 1999, around eighty thousand visitors went to Kedarnath. In 2007, in the high season, Kedarnath saw approximately ten thousand visitors per day, and over the course of the seven-month season about five hundred thousand. Valley spaces in the economic catchment area of the Char Dham Yatra and Hemkunt Sahib became spaces primarily oriented around the *yatra* tourism of middle-class pilgrims, whose expectations for comfortable travel and whose sheer numbers far exceeded the long-term carrying capacity of the mountain environment.

This sudden growth of infrastructure and the nature of economic development connected to pilgrimage and tourism were not sustainable and, as we shall see, unfolded in a manner contrary to the founding vision of the state. It had been known in 2013 for decades, if not longer, that infrastructure development in Uttarakhand, as a primarily mountainous region, needed to take the geologically unstable and relatively inaccessible nature of the region into account in assessments of what scholars of tourism and environmentalists might refer to as the "carrying capacity" and "saturation point" of the region. The setting ought to impose an upper limit on how much the industry ought to grow. Pitamber Sharma (2000, 151), in his study of Badrinath, claimed that prior to the late 1990s pilgrimage tourism did not significantly upset the balance of the Uttarakhand mountain environment and economy because the number of visitors willing to make the difficult trip and their willingness to endure hardship in the name of "religious merit" fit harmoniously with the environmentally aware and economically diversified and localized nature of *pahari* life, and Jagdish Kaur (1985, 189–90) offered a similar argument about how "new tourism" in Garhwal was moving away from the ideal of "austerity" and toward the ideal of "luxury," thereby placing greater strain on regional resources.

A SENSE OF REGION

As Andrea Pinkney has noted in her careful study of recent *mahatmya* pamphlet texts, the second half of the twentieth century also saw the self-understandings of the region that would become Uttarakhand change significantly in ways that influenced depictions of the Uttarakhand Char Dham. After Independence in 1947 there began to be a greater sense of connection to the world of the North Indian plains. In 1955, Pinkney (2013a, 235) observes, an Uttarakhand Char Dham *mahatmya* pamphlet had to call its pilgrimage circuit the "little" Char Dham because it was not well known in comparison to the all-India Char Dham. This sense was immediately complicated in the 1960s by the Sino-Indian War of 1962, which "led the Indian government to target Uttarakhand for intensive infrastructure development, in order to create a defensible buffer zone between the Chinese border and India's plains" (Pinkney 2013, 234–35). Pinkney observes that in the 1960s the Uttarakhand Char Dham was still primarily understood as a paired set of two *dhams*: Yamunotri-Gangotri and Kedarnath-Badrinath. In the Kedarnath valley the next several decades saw the extension of the road from Guptkashi ultimately to Gaurikund, and the territory of the Uttarakhand Char Dham as a whole saw much-increased development of infrastructure and roads and the coalescence of the Char Dham into a single "network of four pilgrimage sites" (239). As the records of visitor numbers maintained by the Samiti attest, these changes in infrastructure clearly correlated with the beginning of a rapid rise in the number of visitors to the region.[10] In the 1980s, Pinkney notes, pamphlets were being offered in a range of languages beyond Hindi—"English, Bengali, Gujarati, Malayalam, Marathi, Nepali, Tamil, and Telugu"—a development attesting to the growing diversity of visitors to the region (242). Further, the 1980s and 1990s saw a modernization of the practice of *yatra* and the expanding overlap between *yatra* and the description and promotion of "places to visit with no religious association." The famous hill stations of Mussoorie and Nainital are good examples of tourist destinations in Uttarakhand not primarily known for their religious significance. These changes made the Uttarakhand Char Dham into a single religious "circuit" that "increasingly integrated into a pan-Indian narrative of commercialized sacred space" connected to modern major urban centers (252, 254–55).

The single circuit of what can appropriately be called *yatra* tourism came to be connected to the powerful brand identity of the region as a whole: Dev Bhumi, Land of the Gods.[11] And at the same time, the label Dev Bhumi referenced the network of local Himalayan deities bound to the landscape who claimed rights over trees, the goddesses whose powerful presences were embedded in earth and water and efficaciously intervened in daily life. This meant that in the "regional modernity" under way in Garhwal broad processes of commodification were blending into place-based networks of divine power that preceded those processes, with the push for the blending being partially provided by *yatra* tourism.

Yatra tourism would then come to function as an important signifier both for a set of internal changes about how Garhwalis were relating to (and shaping) their own deity-pervaded natural landscapes and for a set of outward-looking changes as connections to the North Indian plains, along with a greater sense of cosmopolitanism and linkage to the projects of the modern-nation state, were proliferating.

It now may be seen how this broader story bears directly on our historical understanding of Kedarnath in a distinctive way. The same patterns of regulation, management, and extraction that built roads, cut trees, and constructed hydro-electric dams also reshaped cultural terrain. The premodern religious and political landscapes of Himalayan and Shaiva and Shakta sacred geography contentiously ruled by local Himalayan kings joined the broader life of the subcontinent to a greater extent as they were brought under British political, economic, and cultural influence. Independence in 1947 added new nationalist layers, and the processes leading up to statehood in 2000 formalized a sense of regional *pahari* identity. State involvement (in both colonial and postcolonial settings) with the control and imagining of the natural world (the extraction of natural resources, the building of roads, the damming of rivers, the construction of hill stations) became slowly bound up with the practices of *yatra* tourism in the late nineteenth and twentieth centuries. The mingling of all of these different forces accelerated in the 1990s with explosive results in the early twenty-first century amid the floods, landslides, and earthquakes that were part of the processes of the Himalayan environment, processes whose relationships to human life in the Himalaya were themselves changing in the always emergent production of landscape. I began visiting Kedarnath during this charged time.

THE SEASON

When I was living in Ukhimath and preparing to live in Kedarnath during my dissertation fieldwork in 2006–8 I sometimes heard the season referred to as a six-month "fair" (Hindi: *mela*). The season is immediately preceded by a series of *melas* in the Kedarnath valley that occur in the month of Baisakh: Old (Hindi: Budha) Madmaheshvar, Taltoli Devi (near Lamgaundi), Phegu Devi (near Lamgaundi), Tyudi (Tyudi village near Fata), Jakh (near Jakh Dhar), Maykhanda (Maykhanda), and a small *mela* in Triyuginarayan that is not the main *mela* of the year there. *Melas* in the Kedarnath valley function as some of the most important and fun social events of the year and used to be one of the main times when family and friends who did not live within walking distance would see each other. Gods and goddesses usually make an appearance, but the nature of that appearance varies widely. Taltoli Devi comes out of her temple in a palanquin (Hindi: *doli*) and "dances." Jakh possesses his human vehicle, who, after a period of intense purification and preparation, runs across hot coal and burning ash. In Maykhanda bards sing local versions of the *Ramayana* and narrate a time, centuries past, when a Gujarati prince came to Kedarnath on *yatra* and tried (unsuccessfully) to steal sheep and goats belonging to locals. Later in the day, local deities become present in their human vehicles to receive worship and answer questions. There is also a great deal of commerce involved in a *mela,* particularly involving the sale of sweets, snacks, and toys. *Melas* are by definition short-term events that are, relative to a normal day, a whirl of socializing, consumption, the viewing of traditional forms of performance and entertainment, and interaction with different forms of divine presence. To call the pilgrimage season in Kedarnath a *mela,* then, is to acknowledge with dry humor the many ways in which Kedarnath felt like a larger-than-life place. Let us visit the *mela.*

4

The Season

In this chapter I want to walk the reader through Kedarnath as it felt in 2007 and 2008. As I do so I direct the eye and imagination of the reader in several specific directions. First, I emphasize how diverse experiences of Kedarnath were consistently produced through eco-social engagements with the place as a whole rather than solely focused on a single location there, such as the temple. Second, I illustrate how Kedarnath functioned as a point of overlap for different forms of transregional Hindu pilgrimage and tourism as well as region-specific and some-times valley-specific Garhwali matters. Third, I foreshadow events and trends that, in post-2013 hindsight, were revealed as critical vectors. Kedarnath was in the process of recent and rapid change because of the growing numbers of *yatris* and the pace of recent commercial development. The interdependent levels of energy, devotion, capital, anxiety, environmental stress, and debt related to Kedarnath were rising. There was a sense of a rapidly expanding and swollen bubble.

I begin, after a brief sketch of the rhythms of the season, with extended pre-sentations of two examples that highlight the ways that different locations in and around Kedarnath village served as focal points for human and divine attention: the visit of a local Kedarnath valley deity who wandered over the entire Kedarnath end-valley (thereby introducing the reader to the layout of Kedarnath and the surrounding locale) and the distinctive aspects of the ritual of the ghee massage (Hindi: *ghee malish*) of the *linga* performed by devotees in the inner sanctum of the temple. I then focus on vignettes that illustrate themes necessary for under-standing the Kedarnath of 2007 and 2008: the intensity of the high season, *bhakti*, weather, labor, livelihood, nature tourism, trekking, Garhwali nature spirits, and the relationship of divine *shakti* to the natural Himalayan landscape. I conclude by recalling events that in retrospect felt like previews, foreshocks as I call them, of what would happen in 2013. Pervasive in all of these moments of description is my commitment to attend to the experiential weight of the underlying eco-social interconnectedness of these different times, practices, experiences, modalities, forces, and organic and inorganic entities.

THE TIMINGS OF THE SEASON

An advance party of Samiti, public works, police, and military employees proceeded to Kedarnath several weeks before the season began. They confirmed that the path was passable and that enough snow had melted to allow access to the buildings. Friends told me that there had been years when they would use the not-yet-electrified electric lines as handrails because the snow was so high when they made the preseason ascent. Phone and electrical wires had to be reinstalled and pipes for water reinserted into the ground that had been removed at the end of the previous season. The season began, according to custom, with the three-day procession from Ukhimath to Kedarnath of the traveling form (Hindi: *utsav murti*) of Kedarnath-Shiva, carried in a palanquin (Hindi/Garhwali: *doli*). The departure date of the procession varies for astrological reasons but is usually near or on the third lunar day of the waxing fortnight of the month of Baisakh (Hindi: *akshay tritiya* of the *shukla paksh* of Baisakh). In 2007, the Kedarnath *doli* left Ukhimath on April 27 and arrived on April 30, with the doors of the temple opening the next morning. Bhairavnath was worshipped in Ukhimath the evening before departure and sent ahead to prepare the way and ensure a safe procession and safe opening. Bhupendra and I walked with the *doli*.

The arrival of the *doli* in Kedarnath is the key that activates the energy of this place that has lain quiescent during the winter months. Once the doors are open and Bhairavnath has been officially worshipped in Kedarnath in his Bhukund Bhairavnath form, the season begins. During the first approximately six weeks of the season in 2007 and 2008, in the high season, the area was bursting at the seams, and the Gaurikund-Kedarnath footpath became a continuous stream of *yatris*, often with over ten thousand visitors passing through the place each day. The exact duration of the high season changes from year to year because of the permutation of several factors: the date the temple opens, the beginning of school holidays and vacations for salaried employees, and the onset of the monsoon. In a "normal" year the monsoon will not begin in Garhwal until sometime in July. Once the monsoon begins in the mountains it is not a good time to travel from the plains—the chance of landslides affecting road travel is too great and the plains have started to cool off. In 2007 the high season was perhaps a bit longer than most; one contributing factor was that Jyet (May-June), the month following Baisakh, was longer than usual in 2007. During the high season at Kedarnath, everything is more expensive and

FIGURE 7. The Kedarnath *doli* procession leaves Ukhimath in 2011.

FIGURE 8. The Kedarnath *doli* procession on the way to Kedarnath in 2011.

people are in a hurry. This is also when, correspondingly, the resident population of Kedarnath is highest. Many *tirth purohits* will come for this period and then depart, for various reasons (e.g., a schoolteacher who is also on his holiday).

The high season in 2007 ended around the end of June and the monsoon began. Monsoon season in the Kedarnath valley means that the roads will inevitably be blocked by landslides at least several times; in Kedarnath these occasions were apparent because the flow of *yatris* would suddenly decrease. The end of June and July and the first part of August (the months of Ashadh and Shravan) were the slowest times in Kedarnath; many locals who did stay for the entire season took the opportunity for a quick trip home then. The month of Shravan, Shiva's month, began in 2007 in the middle of July. The time of the monsoon is often when Garhwalis will themselves go on *yatra* within Garhwal. This was also when herders began to arrive in the area around Kedarnath with sheep, goat, and buffalo herds; before then only powdered milk was usually available. Bhukund Bhairavnath was the focus of devotion during this time for locals. Shravan was when Kedarnath valley locals who worked between Gaurikund and Kedarnath themselves came to Kedarnath, first entering the temple and then proceeding up to Bhukund Bhairavnath to offer him oil and thick flatbread cooked with cane sugar/molasses water (Garhwali: *rot*; Hindi: *roti*).

In the eight to ten years preceding 2013, during the latter half of Garhwali Shravan, *kavad* groups began coming to Kedarnath (usually from Gangotri) as an extension of the non-Garhwali *kavad* festival that centered on doing a barefoot *yatra*, bearing pots of water, known as *kavad*, to Haridwar and the Neelkanth Mahadev temple near Rishikesh and that involved millions of devotees per year. Several *tirth purohits* also told me that farmers would often come on *yatra* during the monsoon (or more technically, when the monsoon was happening for them), because little agricultural work could be done during that time. In stark contrast to the intense crowding of the high season, during July and August (and even later) at Kedarnath in 2007 it was possible to spend hours at a time in the inner sanctum of the temple. There was time for locals to play cricket on the floor of a side valley behind Bhukund Bhairavnath to the east.

Just after the beginning of Bhadon in mid-August of 2007, the Samiti sponsored an eleven-day recitation of the Sanskrit text of the *Shiva Purana*, beginning on August 17. Following close on the heels of the ending of the *Shiva Purana* in 2007 was the local festival of Bhatuj, as far as I know celebrated under this name only at Kedarnath and a few other temples in the Kedarnath valley area. Preparations began on August 31, and the final elements of Bhatuj worship were done in the early morning on September 1. Bhatuj is a local form of the common new grain festival known in Hindi as *annakut*. Kedarnath valley locals, especially teenagers, descended on Kedarnath in crowds for this festival, and it felt to me like a *mela*, although I did not hear anyone use this word. Specially chosen *tirth purohits* from the five villages of Ukhimath prepared a large quantity of rice that they then shaped into bricks and placed on top of the *linga* in the temple to form a pyramidal shape. The temple then opened early in the morning for this special *darshan*.[1] There was also a *mela* at Triyugi Narayan in September, one of the most important in the region.

By early September the monsoon had given way to about a month and a half of beautiful clear weather, the best time of the year for walking in the Garhwal Himalaya. Late September and October saw greater and greater numbers of Bengalis in Kedarnath who were taking advantage of the long holidays afforded by Durga Puja in West Bengal to come to the Himalaya.[2] As October advanced the weather began to turn cold and locals began to leave except for the core group whose custom it was to see out the entire season. When Bhupendra and I briefly left Kedarnath to visit Madmaheshwar the only other visitors in Madmaheshwar at the time were Bengalis and a local village on *yatra* with their *devta* (deity).

The season in Kedarnath ends on Bhaiya Duj (Bhai Dooj), a festival that falls on the last day of Diwali. The Kedarnath *doli* typically leaves in the morning and takes two days and three nights to return to Ukhimath. On the penultimate day, the distinguished political and religious figures and devotees whose custom it was to attend the opening and closing of the temple arrived. The ranks of police and military grew as they did what was necessary to secure Kedarnath for the winter. On the final morning, the military ran its food kitchen for all those still remaining in the village. The night before departure, Kedarnath-Shiva received *samadhi puja,* an elaborate procedure that involved the covering of the *linga* with numerous layers of *bilva* leaves, ash, and fruit, as if tucking Shiva in for the winter. In the morning, the Samiti prepared the *doli* for departure. It emerged from the temple around eight in the morning, and representatives of the Samiti locked the temple. The drums and bagpipes of the Garhwali military band sounded, along with the Garhwali drums that accompany the deity on deity processions in Garhwal. Snow had begun to fall in Kedarnath several days before the end of the season. Combined with the sunshine of the morning, this meant that the first several kilometers of the journey were over a footpath covered with melting ice. With the departure of the *doli* from Kedarnath at the end of the season, the area, even more so than during the season, officially became the province of Bhairavnath. Decades ago there was a famous renunciant named Phalahari Baba, memorialized at several different locations in Kedarnath, who reportedly used to spend the entire winter inside a single building, but in 2007 to the best of my knowledge no humans spent the winter. During the first five years after 2013 the offseason has been different and nontraditional: construction workers and heavy machinery have often stayed on in Kedarnath during much of the winter so that reconstruction could proceed and Kedarnath could be as ready as possible for the next high season.

THE VISIT OF NALA DEVI

The visits of Garhwali deities in their palanquins have been an important feature of the season in Kedarnath, although this would not necessarily have been apparent to short-term visitors. Most of the time Kedarnath feels like a supraregional and Pan-Hindu (to use Surinder Bhardwaj's classification) North Indian *tirtha* located in the Himalaya (Bhardwaj 1973). The lingua franca of Kedarnath is Hindi, with many *tirth purohit*s also speaking a bit of the different languages or dialects of their specific patron communities (Gujarati, Punjabi, Sindhi, Bhojpuri, Marathi, Marwari, etc.). Hindi devotional songs and Sanskrit verses play over the temple loudspeakers, occasionally interspersed with the words in Hindi of famous reciters and expounders (Hindi: *kathavacak*) of *Puranas* and the *Ramayana*. Video discs of the *Shiva Purana*, the Char Dham Yatra, the Panch Kedar, and the twelve *jyotirlinga* can be heard when there is electricity, and most of the pamphlets and literature are written in Hindi, though one can also find Gujarati, Bengali, Tamil, Telugu, and Nepali. In the main, Kedarnath is more a Hindi place than a Garhwali place. The *yatras* of Garhwali deities, known as *deora* or *devra* (short for *devta-yatra* or "deity-*yatra*"), are certain moments in the season when in contrast Kedarnath becomes an intensely Garhwali place, a place animated by (to reference again Bhardwaj's taxonomy) regional, subregional, and sometimes even local concerns. With this general framing in mind I now describe the *deora* of Nala Devi (also known as Lalita Mai) to Kedarnath on August 29–30, 2007. In the context of this chapter a description of her visit also functions as an introductory tour of Kedarnath and the Kedarnath-end valley as conducted by a local guide that shows how for some the power and importance of Kedarnath spread out beyond the temple and across the Kedarnath valley as a whole.

As with many *deoras,* we could see the Nala Devi procession and hear the sound of her accompanying *dhol* and *damaun* drums before she actually came over the bridge.[3] Nala Devi came through the horse area and over the bridge marking entry into Kedarnath proper a bit before five in the afternoon on August 29. She traveled as many Garhwali deities travel: as a collection of *murtis* (Hindi: material form of a deity, in Garhwal often found in the form of a metal mask) inside a *doli* suspended between two poles resting on the shoulders of two or four barefooted men.[4] Nala Devi, by which I mean the *doli* containing the forms of her material presence, was swathed in numerous textile adornments, garlands, and dozens of scarves.

Much of the resident population of Kedarnath had turned out to welcome her as she crossed the bridge. Officials of the Kedar Tirth Purohit Association and members of the Samiti welcomed her with garlands and with the chanting of mantras. As with many Garhwali deities on the move, her *doli* was mobile in three dimensions, bending and shaking and bowing. The movement of those who have the *doli* on the shoulders is frequently, when the deity is in such an active mode, jerky and somewhat unstable, as if she is pulling them or forcing them to go in a particular direction. The Kedarnath *doli,* by contrast, does not move in this way, and its progress is statelier. Nala Devi proceeded with this kind of motion down the steps to the small *ghat* (Hindi: stepped bathing area) on the bank of the Mandakini river. She bowed repeatedly to the river until four or five people began to bathe her by dipping vessels in the water and pouring them over the *doli.*

She then came up through the main bazaar, thronged on all sides with people. Local women were throwing flowers and rice, as they do on many auspicious occasions when a *devta* is present, and some were tying on scarves to the *doli.* People were shouting "Jai Kedarnath, Jay Ganga, Jay Lalita Mai." She turned into the lane for the Udak Kund, the enclosed water tank whose water *yatris* are often told is especially efficacious for the destroying of negative karma, and made reverence at the small adjoining Navadurga Devi temple. The Navadurga temple is usually a devotional focus only during specific, Devi-related times such as the festival of Navaratri. With the exception of her powerful river-forms, Kedarnath-based forms of Devi do not, with the exception of Navaratri and the occasional devotional attentions of someone who is a particularly ardent devotee of the Goddess, attract a significant amount of human ritual attention. Devi is much more prominently present in her form of Gauri-Parvati, in Gaurikund, at the trail-head for the ascent to Kedarnath. However, numerous Garhwali *devis,* or forms of the Goddess (along with other deities), visit Kedarnath on *deora.* Nala Devi then went back to the main road of the bazaar and proceeded into the courtyard of the temple. Nala Devi possessed her designated vehicle (Garhwali: *naur*), who made motions of request with his hands and body and was given water.

Thus attended with devotion, spectacle, and fanfare, Nala Devi entered the courtyard, performed one *parikrama* (Hindi: circumambulation) of the temple, and after bowing to the Ganesha who guards the door of the Kedarnath temple, was taken off the shoulders of her bearers and placed in front of the still-closed door of the temple, where she received a small *puja.* When the temple opened for *shringar darshan,* Nala Devi was picked up and entered into the inner sanctum. She, in the form of the *murtis* carried inside the *doli,* was lifted from her *doli* and placed onto a niche in the back wall of the inner sanctum immediately behind the *linga.* Nala Devi would remain there until the next morning when it was time for *puja.*

There is a theatrical aspect to the arrival of Garhwali deities in Kedarnath, and specifically in the temple courtyard. The *devta* becomes more active as it approaches the temple, and the drumming quickens. Often the *deora yatris* begin to dance one of the circular dances that are often claimed as a hallmark of Garhwali culture, and frequently several people (women and men) become possessed. I was able to observe such moments in 2007 perhaps a dozen times. When a deity enters the courtyard, hundreds of locals stop their Kedarnath-related business with *yatris*

FIGURE 9. A Garhwali deity enters the temple along with her insignia and her fellow human *yatris* in 2007.

from the plains and gather for a variation of the Garhwali *mela* where they watch the *deora yatris* and the *devta* dancing/being caused to dance (Hindi: *devta nacna, devta nacana*). As when such possession happens in more typical village and town settings, everyone present becomes particularly focused and serious, extremely engaged with what is happening and what is becoming present. *Yatris* from the plains are frequently unable to interpret what is happening at such a time. Some assume that it is in fact Kedarnath-Shiva who has come into the courtyard. Others do not register immediately that it is a *devta*. Still others recognize it is as a *devta*, take *darshan*, and pay respect without necessarily understanding how exactly this particular *devta* has come to be where it is. A minority of *yatris* will ask a local to explain to them what is going on.

That evening I went to the *dharamshala* where many of the Nala Devi *yatris* were being hosted. Such an arrangement is, incidentally, a demanding situation for the *tirth purohit*. Traditionally speaking, the role of the *tirth purohit* in Kedarnath involves, in addition to carrying out all the ritual needs of his patron (Hindi: *yajman*), finding the patron lodgings and arranging meals. He is a host. However, unlike a lodge owner (there are in Kedarnath both *dharamshalas* and lodges), who may discuss the charges for services rendered up front, the *tirth purohit* cannot. He provides hospitality and performs *puja* and then cannot insist on the amount of the fee he will receive. Thus part of being a *tirth purohit* in Kedarnath is arranging expensive services (food for an entire village, for example) without necessarily knowing what one will receive in return. The situation invokes both traditional values of hospitality and selfless service (Hindi: *seva*) and at the same time is shot through with uncertainty and anxiety because the patrons will not necessarily reciprocate with payment in the appropriate manner when the time comes. When a *tirth purohit*'s Garhwali patrons come, however, I have noticed that they are usually much less worried about payment and view their hospitality actions more through the lens of Garhwali intervillage relations. It is customary for Garhwali villages to welcome *deora* groups with food and lodging as they journey on processions and *yatras* and to expect such hospitality when they themselves are on *deora*.

At one of the *dharamashalas* where the Nala group was being lodged, Bhupendra and I spoke with an elderly man from Nala and one of the older Kedarnath *tirth purohits* in whose *dharamshala*'s canteen we were seated. According to them this was the third time that Nala Devi had come to Kedarnath. The first time had been at least four decades before, and Nala Devi had come not in a *doli* but in a basket carried on someone's back.[5] The Nala man explained that Nala Devi came in order to obtain *shakti*. He went on to link her reasons for coming to his own, saying that there was certainly *shakti* in Kedarnath and that if there was not, then how would he, such an elderly man, have had the power to come to Kedarnath as he did?

Garhwali deities go on *yatra* for various interrelated reasons. One common explanation I heard many times was that Garhwali deities came to Kedarnath to recharge (Hindi: *recharge karne ke liye*): that is, Kedarnath is a charging station for smaller local deities whose "charge" periodically runs out. This was the case with Nala Devi. Another reason, usually much less discussed for obvious reasons, is for the expiation (Hindi: *dosh nivaran*) of a collective problem with supernatural aspects in a particular village or set of villages that the deity has often itself recommended through its oracles. A third reason is simply that the deity wished to go on *deora* and announced its desire by possessing its chosen representative, who then delivered the message, or that particular members of a village or the entire village wanted to go on *yatra* and to take the *devta* with them on the journey (Berreman 1972, 103). Other possible reasons include the habits of a particular deity to go on *deora* once every set number of years or its desire (not limited to going on *deora*) to exercise sociopolitical authority in its community (Jassal 2016). During my time at Kedarnath there were *deoras* of Garhwali deities who came every year, deities who were coming for the first time in forty-five years, and deities who had never previously made the trip. When deities came to Kedarnath, they were worshipped as deities by their devotees but also, relative to Kedarnath-Shiva, they behaved in many respects like a paradigmatic human devotee. A designated human representative (Hindi: *pratiniddhi* or in the Garhwali of the Kedarnath valley: *naur*), often the person whom the deity regularly possessed if a specific person existed, acted on behalf of the deity.

Early the next morning, the *deora yatris* went to the temple and, having done *parikrama*, entered it; to the sound of drums they then brought Nala Devi out into the courtyard. Bhukund Bhairavnath possessed his *naur* as well, a Kedarnath *tirth purohit* from the village of Rudrapur in the Kedar valley. This appearance, while brief, meant that Nala Devi was being welcomed and invited to make herself at home by the guardian of the valley.

After receiving the permission of Bhukund Bhairavnath, Nala Devi began to engage the place beyond the temple and the village. Her itinerary for the day in many ways constituted a map of people, locations, and practices outside the temple that were significant to Kedarnath in 2007. From the courtyard Nala Devi headed out of Kedarnath to the north to bathe in the Saraswati River (which I sometimes heard referred to as the Swargdhari or Swargdvari), a smaller river flowing in from the northeast that joins the Mandakini River about a kilometer behind the Kedarnath village. The government had created a massive barrier wall that began just behind the northwestern edge of Kedarnath village and ran diagonally northeast behind Kedarnath so that the river would not overrun its banks and flood the village. This boundary wall washed out in 2013, and one of the major reconstruction projects was the rechanneling of the Saraswati River back to her pre-2013 course. Nala Devi's bath in the Saraswati (recalling that there was a bathing ghat on the banks of the Mandakini River at the entry to Kedarnath village in which she had already bathed) can be read as an acknowledgment of the numerous sites of power (many of them rivers) lying just north of Kedarnath village that are described in the *Kedarakhanda,* discussed in chapter 2. On the way to bathe, however, Nala Devi stopped briefly at the free food kitchen (Hindi: *langar*) of the renunciant Mahant Chandragiri.

Nala Devi stopped briefly and bowed at the recently built hall of Adi Shankaracarya before continuing to the Saraswati River. Upon arrival at the banks of Saraswati, Nala Devi dipped her canopy/umbrella (Hindi: *chatr*) in the water. Her *doli* was then seated on the banks, and all of the scarves and insignia were taken off, put down, and immersed separately in the river, a procedure that took about an hour. The bathing concluded with a small *puja*. Ganga (Saraswati) water was sprinkled over Nala Devi, and the *dhol* and *damaun* drum players also sprinkled Ganga water over their instruments. With cries of "Jay Ganga, Jay Lalita Maya, Jai Kedarnath," the *doli* was picked up again and, after some discussion of whether to go first to Bhukund Bhairavnath, Nala Devi returned to Kedarnath. She returned by a different path, entering Kedarnath from the northeast, and went to the Samiti compound, stopping by the small shrine where the food offered in the temple was prepared (Hindi: *bhog mandir*). The *Ved-pathis* all came out and prostrated in front of her, and she bowed to them. People then asked for the Samiti officer in charge of the rations, supplies, and preparation of the offerings for the temple (Hindi: *bhandari*). The *bhandari* officer came, and Nala Devi thanked him for providing supplies for her and her *deora yatris*; the group was the guest of the temple committee while in Kedarnath. *Yatris* would sometimes come to this area to have an audience with the Virashaiva *pujari*. Otherwise, this was a primarily a local area.

Nala Devi then reentered the temple for *puja*. Other *yatris* who were in the temple at that time were hastened out as quickly as possible. The *doli* was put down opposite the *linga*, and important local *purohits* and the important members of the *deora yatri* group sat on either side. The *tirth purohits* carried out the puja on behalf of Nala Devi. This *puja,* unlike most of the *pujas* done in the Kedarnath temple, was basically a collective *puja* that stood in for individual family *pujas* done by members of the *deora yatri* group. Thus each family present was instructed to say its name and zodiac sign (Hindi: *rashi*) near the beginning of the *puja* so that they would be included in it. Those wearing a sacred thread took it off to make their vow of ritual obligation (Hindi: *sankalp*) and then put it back on. Nala Devi's insignia were also placed on the *linga* during the *puja*, presumably also for recharging.[6] The inner sanctum was by this time empty of any other *yatris*, though one *tirth purohit* was quietly reciting Vedic mantras in the corner.

Several verses were recited from the Seven Hundred Verses to Durga (Sanskrit: *Durga Saptashati*), and the *puja* proceeded through a fairly normal sequence: Ganesh *puja*, praising of Ganesh, and concentration (Hindi: *dhyan*) on Brahma, Vishnu, Mahesh, Shankar (Shiva), and Kedareshvar (Shiva-in/as-Kedarnath) in turn, along with relevant verses. The *tirth purohit* poured water onto the *linga*, on behalf of Nala Devi, and offered whole rice, applied what on a human devotee would be a forehead mark (Hindi: *tilak*) derived from auspicious substances offered during the *puja* such as *sindhur* powder or sandalwood paste, offered a sacred thread, recited verses in praise of Shiva from texts such as the Hymn of the Greatness of Shiva (Sanskrit: *Shivamahimnastotra*), and performed *arati*.[7] The *deora yatris* were told collectively at various points during the *puja* to bow their head to the *linga*. After the *arati* the *tirth purohit* massaged the *linga* with *ghee* on behalf of Nala Devi. Everyone, *deora yatris* and Nala Devi alike, circumambulated the *linga* before leaving the inner sanctum, stopping on the way out to worship Parvati. Upon emergence from the temple, Nala Devi then circumambulated the temple from the outside and headed up the side of the valley for Bhairavnath *puja*. At this point activity bifurcated. A group of ten to twelve *purohits* remained by the Havan Kund, a covered raised platform on the eastern side of the temple that was both the standard location for large fire offering rituals (Hindi: *yagya* or *havan*) in Kedarnath and the base of operation in the temple courtyard for renunciants. They began a *yagya* sponsored by *tirth purohits* from Rudrapur (the Kedar valley village that is the natal village of Bhukund Bhairav's *naur*). Nala Devi went up to the shrine of Bhukund Bhairavnath, and a brief *puja* was performed to Bhukund Bhairavnath. Many *yatris* make the small trek (approximately fifteen to twenty minutes in each direction) to the shrine of Bhukund Bhairavnath if they have the time and energy to spare. Fewer *yatris,* except for the nature-oriented trekkers who both have the inclination and have planned to include time for a longer stay, proceed to wander about on the high plateau immediately behind the Bhairavnath shrine. Here Nala Devi began again to display a marked desire to move about the broader area of the Kedarnath end-valley.

Nala Devi's actions here illustrate the ways in which significant practices (of both deities and humans) in Kedarnath do not limit themselves to focused inter-actions with specific sites. From the shrine of Bhukund Bhairavnath, Nala Devi began to wander (Hindi: *ghumna*). This wandering distinguished the three local Kedar valley *devis* who come to Kedarnath from the other Garhwali *devtas* who visit. Nala Devi, Phegu Devi, and Rampur Devi all are known to potentially wander over much of the length and breadth of the Kedarnath end-valley. *Ghumna*, or *ghumgam*, is the Hindi word used by locals when they describe a jaunt by foot they have taken that was not for a specific purpose other than to enjoy a walk; in this sense Nala Devi was acting like a Kedarnath valley local in Kedarnath, which she was. She first headed south on the Bhairavnath plateau halfway down the eastern side of the valley to a large rock known by some as "Old Kedar."[8]

From "Old Kedar" Nala Devi came down the hill near the helipad. By this point a cold rain had begun to fall. As Nala Devi started to turn north, the *deora yatris* began trying to convince Nala Devi to stop wandering and head back to the bazaar. Bhupendra and I remembered their conversation as something like the following: "Speak, what should be done, if you need to keep wandering then say so, but in this rain we cannot help you wander any more. Also now it is necessary to go into the bazaar, everyone has invited you. We would like to take you to at least half the places in the bazaar today—the rest is up to you [Hindi: *baki teri marzi*]." This conversation serves as an important reminder of the challenging conditions one finds at Kedarnath. At 3,500 meters of altitude, walking wet and barefoot around the end valley in a cold rain with the temperature only a few degrees above freezing was difficult enough that the *deora yatris*, themselves mountain dwellers (Hindi: *pahari*) were willing to challenge their deity. Given that the satisfaction and beneficence of the deity is a large part of why villages bring their deities to Kedarnath, this is a telling indicator of the physical state of the *deora yatris* at this point.

Nala Devi began to head back to the bazaar. On the way, she stopped at Retas Kund, a pool of bubbling water enclosed by a shrine that (as mentioned in chapter 2) for tantric practitioners possesses important alchemical significance, and as hundreds of thousands of other *yatris* have done, once a sound was made (in this case a conch was blown) the group saw bubbles come up through the water in the pool. She continued on through the center of Kedarnath and returned to the hall of Adi Shankaracharya (i.e., Shankara) on the eastern side of the village in response to the invitation of an important renunciant from Varanasi who bore the title Mahamandaleshwar. The Mahamandaleshwar was carrying out a two-month period of extended devotion, meditation, and puja (Hindi: *anusthan*) in Kedarnath as part of his program to carry out a series of *anusthans* in each of the twelve *jyotirlingas*. The hall of Shankaracarya was a recent structure that housed the ostensibly original location (Hindi: *sthal*) of Shankara's passage into liberation/death (Hindi: *samadhi*) along with a *murti* of Adi Shankaracarya as well as a new, gigantic *linga* made of crystal.

For most of August and September, the hall of Adi Shankaracarya was the site for the public daily *puja* and *atirudra yajna* (an especially elaborate form of the Rudri/Rudrashtadhyayi fire offering ritual that lasted for the entirety of the two months of the *anushtan*) of the Mahamandaleshwar. The Mahamandeleshwar had hundreds of followers visit him during his two months in Kedarnath, especially in the last week, culminating in a celebration of the final offering of the *yajna* (Hindi/Sanskrit: *purnahuti* or "last offering") in which all present made the final offering into the fire together, a celebration that involved conservatively several hundred people. He ventured only rarely to the temple. It was clear in discussion with many of his followers that his presence was the primary motivating factor for their journey.

The Mahamandaleshwar and the Brahmins he employed for his *yagya* (composed of both local *tirth purohits* and Garhwali Brahmins from other parts of Garhwal) worshipped Nala Devi with verses from the Durgasapashati and *svasthi vacana* mantras. Then all the *deora yatris* were given chai and snacks. Nala Devi then left the hall of Shankaracarya and went to the bazaar. In the bazaar,

Nala Devi stopped at almost each shop, where she was garlanded with flowers and scarves, was sometimes worshipped with a small *puja,* was given small gifts of money and rations, and then was asked for blessing (Hindi: *ashirvad*). Several older local women who were visiting Kedarnath for several weeks (wives, aunts, and grandmothers of Kedarnath *tirth purohits*) garlanded her. At this point in her visit her activity was no longer that of a *yatri* in Kedarnath. In the bazaar she was a Garhwali deity on *deora* in a bazaar filled with people from her own area, people to each of whom she owed *darshan* and *ashirvad* and who were bound by the especially high level of reciprocal obligations of offering implied by her status as a local *devi* of the Kedarnath valley.

At this point the *deora yatris* again requested/told Nala Devi to go straight back to her resting place, saying, "We cannot help you to wander any more in this rain." She then returned to the *dharamshala* where she was based and remained there at rest until eight in the evening, when she received *arati*. While she was resting, at about 5:30 p.m., Rampur Devi (the second local *devi* of the season) arrived and was greeted in much the same fashion. The next morning, Nala Devi and her *deora yatris* got ready for departure. Some of the yatris were having their family details recorded in the book of records (Hindi: *bahi*) that *tirth purohits* keep of the visits of their patrons. Nala Devi went and circumambulated the temple before visiting more shops and several lodges on her way out of Kedarnath.

The visit of Nala Devi to Kedarnath marked one of the moments when Kedarnath became a Garhwali place, a place where Garhwali attitudes surrounding the presence of *devtas* on the move superseded the economic aspects of *yatra* and temporarily reconfigured many other normally operative distinctions between locals and *yatris* who were usually from outside the region.[9] Her visit, the visit of a paradigmatic *yatri,* showed that while the temple was the primary focus of her visit she also made it a point to pay her respects to Bhairavnath, to bathe in the river in two different places, and to wander across the land in the midst of challenging weather. She engaged the place as a whole in a way that muddies distinctions between nature and deity, local and visitor, temple and place, human and divine agency.

THE GHEE MASSAGE

The massaging of the *linga* with clarified butter or *ghee* (Hindi: *ghee malish*), on the other hand, enacted a spatial focus that was closely focused on Shiva's presence in the inner sanctum of the temple itself rather than the spatial focus of Nala Devi that was spread out over the entire end-valley.[10] In the Kedarnath of recent decades anyone could, as part of *puja,* touch the *linga* in the inner sanctum of the temple during the time for public *puja* in the morning and massage it with *ghee* with her or his own hands. Anecdotally, this practice appears to have replaced the practice

FIGURE 10. A jar of *ghee* next to several partially prepared *puja* trays.

of embracing the *linga* that some older *tirth purohits* remember from earlier decades and that had been mentioned by earlier visitors to Kedarnath, a practice that would have been easier before the *linga* was enclosed in the metal frame, the state of affairs in 2007. Many Kedarnath locals loudly insisted that Kedarnath was the only place where this *ghee malish* was performed. Many *yatris* informed me that this was not the case, but it is nonetheless the most distinctive practice currently linked to Kedarnath and does stand as a unique practice for many.[11]

At the beginning of the 2008 season, it was still the tradition at Kedarnath that every single person or family could have a seated *puja* in the inner sanctum itself. This full access was often contrasted with the highly limited access to the deity *yatris* typically experience in Badrinath. The *puja* itself was fairly standard and often included an *abhishek* of the *linga* (Hindi: the bathing of a *murti* with auspicious materials). Standard *puja* offerings usually included a sacred thread, camphor, incense, whole dried rice, oil lamps, forehead adornments (Hindi: *bindi*), vermillion powder, nuts, split chickpeas, and raisins, and more expensive *pujas* added scarves, plastic flower garlands, and coconuts. Sometimes *yatris* would bring some of the materials for the *puja* from home. Even the least expensive *puja* tray always came with *ghee*. *Puja* took place in the inner sanctum of the temple around the periphery of the metal frame enclosing the *linga*. This was the general ritual sequence: invocation (Hindi: *avahan*), initial vow (Hindi: *sankalp*), *puja, arati,* and offering of flowers (Hindi: *pushpanjali*), and then *ghee malish*. Each member of the family would take semisolid (because of the air temperature) *ghee* into their hands and was urged to rub/massage the *linga* with *ghee* while the *purohit* recited a mantra:

> I am a wicked person [Sanskrit: *papa*], I am one whose actions generate wickedness, my inner being [Sanskrit: *atma*] is full of wickedness, I am one from whom wickedness arises. Save me, lord of Parvati! O Shiva, you destroy all *papa*.[12]

The meaning of the *mantra* accords with the sense that the end of the *puja* sequence, in the Kedarnath valley, often includes a declaration of the sponsor's unworthiness and a blanket request for forgiveness from the deity.[13]

For many *yatris* this intimate embodied encounter with the *linga* is extremely special, an experience usually unavailable to them. For others it is merely an elaboration of something they already know. There was often the sense that the *ghee malish* derived in some way or other from the paradigmatic actions of the Pandavas. Returning to another part of the discussion with the *tirth purohit* Tiwari-Ji discussed at length in chapter 1 provides an example in which the *ghee malish* was connected specifically to Bhima. When I prompted Tiwari-Ji about the *ghee malish* during our conversation, he responded:

> They say that Bhima hit with a mace at that time, saying, "Where has he gone, all of them have gone." He did not have intelligence, he had strength but not intelligence, it is well known . . . that, regarding someone who has strength . . . he is totally *mindless*

["mindless" said in English]. That's how it happens. . . . He had very little thinking power. . . . He hit God and then pulled him. . . . Therefore, Mother Kunti said, "Son, you made a great mistake." When God [Hindi: *Bhagwan*] gave *darshan* there and when he became invisible again, Mother Kunti said to Bhima, "You should apply a repair . . . to where you hit God with a mace," and there that repair is applied. "I am a wicked person, I am one whose actions generate wickedness, my inner being is full of wickedness, I am one from whom wickedness arises. Save me, lord of Parvati! O Shiva, you destroy all *papa*" [mantra recited in Sanskrit]. Doing this, the repair is applied. . . . For the expiation of wickedness, *ghee* is applied to God.

In the context of a book describing the relationship of the Virashaiva tradition to Kedarnath commissioned by Kedarnath Rawal and Virashaiva *jagatguru* Bheemashankar Ling, Shobha Hiremath (2006) writes of the *ghee malish* in this way: "The most mysterious and astonishing point is when you caress the *linga* form of the Jyothirlinga [*sic*], you can feel the backbone and nerves of the back of the male buffalo, each and everyone irrespective of caste colour and creed is permitted to touch, feel, and express their devotion by smearing butter to the linga as a religious ritual. The souls of the devotees soar high with reverence and ecstasy." Whatever the valence, a ritual personal encounter with the distinctively shaped *linga,* thickened by connections to webs of experience and meaning about Shiva and his paradoxical formed formlessness, his connection to the Himalaya and to the Goddess, and his interactions with Pandavas, has been a central part of visiting or residing in Kedarnath for many.

The *ghee malish,* carefully analyzed, can be seen as a model that maps to broader trends, experiences, and textures found throughout Kedarnath. Specifically, it can serve as an analytic microcosm for understanding the stresses of the high season. To understand how, it is first necessary to briefly review the temple timings. In 2007 and 2008 a normal day in the Kedarnath temple began with the Kedarnath *pujari* offering food (Hindi: *bhog*) to Kedarnath-Shiva at around 4:15 a.m. After this came the time for special *puja* (Hindi: *vishesh puja*), *puja* typically involving the presentation of costlier offerings in greater quantity, for which a ticket had to be purchased from the Samiti. Any *puja* performed during this time required the presence of a Brahmin *ved-pathi* who was a Samiti employee, even if the patrons of the *puja* wished to use their own *tirth purohit* as well. General *darshan* and *puja* were available from 6:00 a.m. until 3:00 p.m. This time was often described as the time of bliss-*darshan* (Hindi: *nirvan darshan*).[14] At this time anyone who waited in the queue could directly touch the *linga* and sponsor a *puja* in which each devotee could massage the *linga* with clarified butter (Hindi: *ghee*) with his or her own hands. This opportunity is not always available in large temples and was often mentioned as a special aspect of visiting Kedarnath.

At three the temple would close and the inner sanctum would be cleaned, followed by the daily fire offering (Hindi: *havan*) and the offering of the second *bhog* of the day, and the *pujari* and his assistant would begin the adornment (Hindi: *shringar*) of the *linga* that preceded the evening temple *puja*. After the second *bhog,* only the *pujari* and his assistants could enter the inner sanctum. The doors opened at 5:00 p.m. for what was often called adornment-*darshan* (Hindi: *shringar darshan*), and then evening *puja* and the offering of light and flame (Hindi: *arati*) typically commenced at 6:00 p.m. and lasted approximately one to two hours. During this time *yatris* passed briefly in front of the door to the hall between the antechamber/public hall (Hindi: *sabha mandap*) and the inner sanctum for a brief *shringar darshan* from afar. Around 8:00 p.m. the temple would close until the next morning. Diana Eck (2012, 231) evocatively characterizes the relationship between the *nirvan darshan* of the morning and the *shringar* and *arati darshan* of the evening: "The rhythm and juxtaposition of these two evoke in ritual the Shaiva theology of a god who is both describable and indescribable, both apprehensible and utterly transcendent."

Evening *arati* in Kedarnath was a special time. After a luminous glimpse of the adorned *linga,* a brief moment obtained amid a press of excited devotees, you exited the temple and circled it in the auspicious direction as evening fell. You then joined the growing crowd of people in the temple courtyard who had already taken *darshan*, a crowd that included *yatris* and many locals who were taking a short break from their duties. People were singing devotional songs (Hindi: *bhajan*) from home. Some lit lamps and placed them on the ground or one of the temple walls or made a donation to one of the *sadhus* whose "duty" it was to stand outside the temple during evening *arati*. More so than during the day, the crowd in the temple courtyard included many of the police stationed in Kedarnath, a reminder that Kedarnath is close to the border of the Tibetan Autonomous region and that the state and national governments pay careful attention to what happens here. At the end of the *arati* the Virashaiva *pujari,* accompanied by his Garhwali Brahmin assistant, both of whom are today employees of the Samiti, strode outside and offered the light-flame of the torches of *arati* into the darkness first in the direction of Bhukund Bhairavnath and then to the assembled crowd. Evening *arati* was usually a time when one felt a bond with others because everyone was in this special place at this special time. It was a moment that illustrated the idea, discussed in the Introduction, that the *shakti* of Kedarnath is either/both inherent in the place and a result of the accumulation of what people have brought with them to Kedarnath over the centuries.

In the high season in 2007 and 2008 these timings changed dramatically. Special *pujas* began as early as 11:00 p.m., preceded by the (technically the next day's) morning *bhog,* and continued until around 6:00 a.m., when the doors would open for general *darshan* and *puja*. *Yatris* waiting for general *darshan* and *puja* often had to wait for two to five hours. In 2007, I never saw the temple close for the afternoon until all *yatris* had been through the temple. This policy effectively pushed back the rest of the day's schedule by several hours. In the third week of the 2008 season, I saw the doors close at least once (to the best of my knowledge for the first time) at 3:00 p.m. before everyone had received *darshan*. In 2007 and 2008, it was not clear that the *ghee malish* would continue in its current form because the ever-increasing numbers of *yatris* present in Kedarnath during the high season were making it difficult to provide every *yatri* with the opportunity for seated *puja*, and there was a great deal of speculation about how the ritual might need to change. These incipient changes in temple ritual illustrate the stresses placed on Kedarnath by the high season.[15]

THE HIGH SEASON

The high season was a time when the complex eco-social system of Kedarnath was bursting at the seams, a high-altitude whirl of religious power, devotion, sightseeing wonder, big business, power struggles, changing weather, and over-crowding. The differing ways that the connotative web of story and ritual described in chapter 1 and chapter 2, blended with the complex imaginations of landscape whose trajectory I described in chapter 3, framed experiences in Kedarnath were on full display in the brief scenes and conversations I describe. There was also a high degree of variation (and overlap) in the spatial setting receiving focus at any given moment: sometimes the temple, sometimes the physical context of the path, sometimes the story-enchanted Himalayan land-scape in a bigger sense, and sometimes the inner landscape of the speaker. Perspectives on Shiva's presence in the place and/or the temple could be seen to vary in these scenes as well.

The first several snapshots I present in this section are from May 13, 2007. This day began in a large *dharamshala*. My conversations usually took place in the entry room, unless the manager conducted us to one of the rooms and asked a group if we could speak to them in their room. I first spoke with a middle-aged woman from Mumbai who was doing the Char Dham Yatra for the first time. She had originally meant to come the day before, but there had been a strike by the horse-drivers against the growing number of helicopter flights, so they had come one day later, causing them to miss a 2:00 a.m. appointment for special *puja*. They had brought all of their own supplies with them for *puja*. They had not heard prior to arrival the story about how the Pandavas came to Kedarnath, but they did know that Shiva had become manifest (Hindi: *prakaṭ hue*) in Kedarnath by his own volition. They had tried to walk from Gaurikund, but after three kilometers they began to have difficulty breathing, so they paid someone to carry them in a palanquin (Hindi: *palkhi*) the rest of the way.

I had already heard about this strike. The day before I had been sitting at one of the tents by the helipad provided by one of the private helicopter companies for the use of customers and employees. At that time, I had spoken with one of the helicopter company managers and a three-generation family of *yatris* from Andhra Pradesh. The father was an industrialist. He said that Kedarnath was a better place to go on vacation and waste money (Hindi: *paise barbad karna*) because not only was the scenery better than that of a hill station but also there was the added benefit of *darshan*. *Yatri* families will often mention that one purpose of a family *yatra* is that it may help to instill traditional values in the children of the family. He was indignant that the horse drivers and porters were striking against the increasing practice of journey to Kedarnath by helicopter. Didn't they understand that more jobs would be created by the economic development resulting from these helicopter *yatras*? The helicopter company manager, a Kedarnath valley man, said that the horse-drivers had become stubborn like their animals and were protesting because they were not receiving any compensation from the helicopter companies. When I pointed out that some people felt that Kedarnath, as a traditional place of pilgrimage, was meant to be difficult to reach, he became testy. Are you trying to tell me, he asked, that my old mother should not have come? She could come only by helicopter.

Later that day, out in the bazaar, I was invited to tea by a policeman whose duty was to keep order in the temple line. He insisted that only 1 to 2 percent of the people who came to Kedarnath were true devotees (Hindi: *bhakt*). The rest had come only to make a perfunctory touch of the *linga*. They were not even willing to endure the pain (Hindi: *kasht*) of waiting in line for a little while or coming on foot. He was also extremely critical of people who could afford to come by helicopter and purchase a ticket that granted them the permission to enter through the side door without waiting in line. He felt that *kasht* was a requirement for *bhakti*. This reminded me of something a Hindi teacher of mine once said about her experience of *puja*. Often during *puja*, she said, people put their *kasht* onto God (Hindi: *Bhagwan*). But what about the *kasht* of God? When she did *puja* she tried to take on God's *kasht*.

Arrival by foot in recent years has carried particular resonances. Pilgrimage by foot (Hindi: *paidal-yatra*) is the iconic form of *yatra* and symbolizes the pain (Hindi: *kasht*) and inner production of energy and focus (Sanskrit: *tapas*; Hindi: *tap*) that the practice of *yatra* generates (Whitmore 2016). Yet in 2007 and 2008 (and later when I visited in 2011), to walk to Kedarnath was to be in a dwindling minority. Far more common was to ride on horseback, to be carried by porters, or to come by helicopter. Helicopters passed overhead several times an hour. If you arrived by helicopter (which I have never done), then you overflew the path, armored by your ability to purchase the ticket but also temporarily cast as an observer of the older, more traditional forms of pilgrimage taking place beneath you. You experienced a wide view of the mountains and the river valley. If you were in a helicopter flying up the Kedarnath valley, then you were flying at almost

FIGURE 11. A helicopter lands in Kedarnath in 2011.

half of the height at which commercial airliners fly at cruising altitude. But, of course, this was somewhat removed from the practice of walking in the footsteps of the Pandavas.

In early afternoon I headed toward the temple courtyard. The queue was long and the police were required to intervene to keep order as quarrels broke out between those patiently waiting in line and those attempting to jump the queue. I did not wait in line but wandered up and down the line and tried to talk to people, with moderate success. Then the weather turned. In nonmonsoon Kedarnath it was common to have strong sunlight and clear weather in the morning and early afternoon, but clouds and precipitation would often arrive in the mid- to late afternoon. You could usually see the cloud wall advance up the valley with measurable speed in the afternoon. I went and sat in a shop owned by a friend of mine that sold religious objects and books. A group of *yatris* from Gujarat were castigating him for not having more material written in Gujarati. Several women wanted to buy cassettes of devotional songs (Hindi: *bhajan*). He responded curtly that he no longer stocked cassettes and that they should purchase a Char Dham video instead.[16]

I went again to a large *dharamshala,* this time for a prearranged conversation with a teacher from Delhi on his nineteenth visit to Kedarnath. I was able to partially record the conversation, which took place primarily in English with frequent excursions into Hindi. He said that he had been to all of the *jyotirlingas* at least sixteen times and that he had sponsored and carried out twelve *maharudrams,* a *maharudra* being over one thousand recitations of the *Sri Rudram* Vedic chant (Sanskrit: *Rudrashthadhyayi*). He mentioned a pilgrimage system known as the seven Kailasa that included Manimahesh in Himachal Pradesh and Kedarnath. He contrasted India as a place of mystery with America, where people are interested in demystifying. He said that in India science and tradition are integrated and gave the example that anointing the forehead with sandalwood paste (Hindi: *chandan tilak*), commonly done after some forms of *puja,* releases endorphins that give you pleasure. When I asked him about the shape of the *linga* he said that it was the shape of the Himalaya. He told a version of the Pandava story in which he said that Shiva initially had been aloof from the Pandavas because it was not yet the proper time for their *darshan.* When Shiva took the form of a buffalo Bhima perceived it and stood his legs on two hillocks. Shiva, as Lord, could not walk between Bhima's legs. At this point Shiva realized that Bhima had understood his disguise, so he went into the earth. The hump of Shiva remained in Kedarnath and the Pandavas went to heaven. His navel became established in Madmaheshwar, his matted hair in Rudranath. Ravana, the archvillain of the *Ramayana,* had actually visited Chopta, the tourist town on Ukhimath-Joshimath road that lies just below Tungnath, another of the Five Kedar shrines. The teacher also mentioned "Old Kedar" (Hindi: Budha Kedar), a Shiva temple near Uttarkashi connected to the Nath tradition but not often connected to the Five Kedar shrines. He also mentioned Pashupatinath in Nepal and said that *darshan* of Pashupatinath was necessary for getting the benefits, or fruit (Hindi: *phal*) of the journey to the different *jyotirlingas*.

When I asked him about the *ghee malish*, he said that the reason for it was practical. It was difficult to grow flowers near Kedarnath, whereas it was easy to offer *ghee*. I asked him why he kept coming to Kedarnath. He said that every time it was different and unique. For example, this time was the first time the weather was clear. Once he came in October and there was a heavy snowfall. They had no clothing. He used to come half-naked (recently he has started bringing gear). Once he came when it was raining and everything got soaked. Once he came in October with nothing. Today's worship had a different vision, a different feeling. He started to talk about a renunciant who had been staying (reputedly continuously) for twenty-seven years in the Ramanandi ashram built into the mountainside at Garud Chatti several kilometers south of Kedarnath on the path. He said that this renunciant had kept very quiet until the last two or three years and had finally begun to share. Sharing experiences was good. Always something was different. This teacher always journeyed to Kedarnath on foot, even though he used to have varicose problems. Now he had diastolic dysfunction (a swollen heart) and it took him twelve to fourteen hours to walk up, but still he came.

When I asked him about experiences in the inner sanctum he had this to say:

> Sometimes you get . . . feelings. Sometimes you get visions, sometimes you see something. [I asked for an example] . . . I don't know, I don't remember, but it is there. Some visions come, some noise comes sometimes. Some days we are with ourselves only, we just keep on chanting. . . . Sometimes you feel smells . . . sometimes you feel some . . . vibrations . . . after that one should not say all these things because one does not know what it is exactly. Generally, they are saying you should not share such things but since you are asking . . .

He himself sometimes got a sort of ticklish feeling, sometimes felt as if he were naked and would cry. There was a beginning, a straining toward a beginning. He said that one of the problems with sharing experiences was that there were dangers. Development on a spiritual path could be very dangerous, as could sharing. Each stage felt like the ultimate. I asked him if he felt that Shiva possessed name, form, and attributes that could be expressed through language (Hindi: *sakar*, also termed *sagun*). He said Shiva ultimately was both formed and formless. If you worshipped Shiva as *sakar* then you saw a material form. The experience was no doubt *sakar*, but it was so powerful that you could not bear the full *sakar* form. For example, the full form of Lord Ganesha was as powerful as millions of suns. We could not bear even one form. Thus we used idols (his term) and *lingas*. The problem was that people got stuck. You had to go ahead and keep progressing as if you were progressing through a system of education. There was no difference between Islam and Hinduism except that one started with the formless (Hindi: *nirakar*, the opposite of *sakar*). You had to be very powerful to perceive the *nirakar*. You had to start with form because formlessness was too hard.

He related what he told his students because he often brought students with him on his pilgrimages. He used particular events and experiences as justifications for worship of Shiva. It was because of the protection of prayer, for example, that a train journey had been successful in a time of terrorism. Last year on a trip to Madmaheshwar they had left Rhansi at five in the afternoon. He had injured his legs and the group had gotten lost. Twenty-one students were on one the wrong side of the river valley. People were crying. Suddenly someone appeared and guided them. If it was not a miracle, then it was an amazing coincidence. Another time he was traveling to Kedarnath and he had an oxygen problem in the fog. Someone came and guided them up to the bridge and then vanished. He explained that by "vanished" it meant that the mysterious person suddenly went away quickly and they searched and could not find anyone. Students also had mentioned an approximately eighty-year-old man who helped them once and then disappeared too quickly for someone of his age. Sometimes local people said that they heard voices at night sometimes, for example in Garud Chatti. The teacher mentioned the oft-related story about Badrinath that in places like Badrinath and Kedarnath for six months the gods came and worshipped here. He once had seen a man on the path who was moving but without walking. He thought it must have been the deity Kal Bhairav (Bhairavnath) or some semidivine and/or monstrous member of a troop (Hindi: *gana*) of Shiva's followers or some ancient saint. He mentioned several similar occurrences in and around Badrinath that involved sightings of Uddhava, a friend of the god Krishna.

We began to discuss the ways that Kedarnath had recently been changing. The teacher said that the essence of Kedarnath would not change because of recent development but that there would be so many layers over that essence that people would not be able to see the true character of the place. He likened this problem to the ways that the seven sheaths (Sanskrit: *kosha*) of bodily material prevented us from directly perceiving the life force (Sanskrit: *prana*). Walking to Kedarnath barefoot was better for getting the full experience of the place, an experience that involved both pain and pleasure, but of course not everyone was able to do this. It would be good if the nearby glacial lakes of Gandhi Sarovar and Vasuki Tal became more accessible, but this might spoil their atmosphere. People were beginning to be trapped by a strict *yatra* schedule, which meant that they were more likely to just go to Badrinath or to rush through the Uttarakhand Char Dham Yatra.

At the same time, even in Mumbai intense *bhakti* was happening for lots of people. Coming here was not strictly necessary, but for certain kinds of religious experience one had to wander like the saints. There must be a purification process, the banana must be unpeeled. He said that I should be like the chiku seed and not like the mango seed. The chiku seed was totally dry and could be cast aside, whereas the mango seed always seemed to have a bit more juice and therefore could not be cast aside. Being in the world was good. He told students: get involved but not attached. You have to taste the world. As the conversation ended, he said he hoped that he had not wasted my time. When people like me came, unfortunately most people like him were not able to explain adequately. Most of his students did not learn through his explanations; instead he had to create the conditions for them to have learning experiences of certain kinds because he was interested in their salvation. He gave the example that he had once caused his students to perform an *abhishek* with seven liters of milk. They had found this wasteful at the time and later they understood. Most of his students were elites who liked Western things and did not believe in traditions but rather thought of them as myths, so their minds had to be changed slowly. At the end of the conversation he gave me a laminated AUM/OM image for my own use.

Finally, Bhupendra and I visited a different large *dharamshala*. There a Kedarnath valley local friend who worked for a national *yatra* company introduced us to a group of about twenty-five *yatris* from Gujarat as they were eating their dinner (provided by the *yatra* company) in the canteen of the *dharamshala*. It was clear that they were having a blast. I stood in front of them and introduced myself and a rowdy question-and-answer session followed. Members of the group felt that the pursuit of the Pandavas was appropriate because they were in pursuit of *mukti*. They said that Kedarnath was Shiva's dwelling place (Hindi: *nivas*) but that Shiva was not Kedarnath. Many people were hearing the story of the Pandavas' journey to Kedarnath for the first time, but one person said she had read something about Kedarnath in the well-known comic book series *Amar Chitra Katha*. We discussed the shape of the *linga*. I said to me it looked like a mountain. They said one stone is just a stone but this other stone is Shiva, so we worship it as Shiva even though to you it looks like a stone in the shape of a mountain. To us it looks like the back of Shiva or the hump of his buffalo-form. Bhupendra noted that they themselves said they should be seen as both pilgrims and tourists. One *yatri* said that at the time of *puja* there is utter concentration/meditation (Hindi: *dhyan*). When you look at nature, you take its full joy. After that, Bhupendra and I happily had dinner with the group and then went back to the room we shared for the night. There were many days like this.

WEATHER

The weather is one of the most important constituent factors in how people experience Kedarnath.[17] The weather in Kedarnath is distinct, even different from the kinds of weather found on most of the path from Gaurikund. It is colder and there is often more fog and precipitation. It snows and hails sometimes, even in the summer. Preparing to visit the temple and waiting in the queue (to say nothing of making the trip from Gaurikund) when the sun is shining and one can gaze on the glaciers spread out behind the temple is radically different from waiting with bare feet and usually inadequate rain gear and warm clothing when fog, rain, hail, and sometimes snow descend. Once when I was conversing with a *yatra* guide from Varanasi as we walked back down to Gaurikund, he confessed that he had not bathed before entering the temple because it was so cold that he simply could not make himself do it. He was comforted by the well-known local idea of the "wind bath" (Hindi: *hawa snan*). This phrase refers to the idea that the meteorological environment purifies those who enter Kedarnath in a way that that makes actual bathing optional. I heard locals refer to this idea many times. It was an implicit acknowledgment of the difficulties inherent in a high place where the balance of wind, sun, and precipitation was always changing. As with the previous examples, this idea shows how the natural environment exercises a religious power in its own right that is both connected to what is present in the temple (one should be pure to enter the temple) and different from it (the purificatory power arises from the weather and not, in a direct sense, from the presence of Shiva). Since 2013 there has of course been another layer to all of this: the overshadowing memories of 2013 triggered by bad weather.

TENSION

The excitement of the high season in 2007 and 2008 had many different valences. On the one hand, it functioned as a religious spectacle that celebrated, as Shukla-Ji described in the Introduction, the *shakti* of the god and the place, and the traditional and at the same time modern power of *yatra*. On the other hand, the number of people in the place and the amount of money circulation created stressful situations. On May 27, 2007 there was a confrontation between a *sadhu* (Hindi: renunciant, ascetic) who had been living in the Kedarnath valley for decades and a group of *sadhus* who he felt were not behaving with proper restraint (Hindi: *maryad*). It escalated into a physical confrontation that settled down only after the police became involved. The angry *sadhu* said that some of the resident *sadhus* were bothering pious *yatris* who were coming to Kedarnath for *darshan* and perhaps even stealing from them. These false renunciants, he said, were hoarding their donations for their own use. Instead, he felt, they should be using those funds to give *yatris* chai to drink and food to eat, and to buy fuel for fires to warm the cold, shivering *yatris* who found themselves unprepared for the high mountain weather. He said that these false renunciants were making the temple dirty. *Yatris* were dying from cold, he said, and these *sadhus* didn't care.

In June a well-known politician visited Kedarnath. As is often the case when such figures visit, he was received by representatives of the Kedarnath Tirth Purohit Association. On his way to the temple the politician was asked to lend his support to the extension of the motor road up to Kedarnath. He said that Kedarnath was a holy place and shouldn't be turned into Switzerland. He pointed to how Gangotri (the second of the Four Dhams, near the source of the Ganges River) had become a honeymoon spot, and he referenced a time when his audience's forefathers had had to work in an environment where visitors came up from Rishikesh on foot. After *darshan* and *puja* in the temple the conversation continued. The politician said that the government was looking into the impact the number of visitors was having on the rate of glacial melt and that government experts were now considering what sort of development (Hindi: *vikas*) should happen so that more harm did not come to the environment.

On that day the line was very long, and even though it was well into the afternoon the temple had not yet closed for *bhog*. There was an uncomfortable amount of tension and confrontation in the temple courtyard between *tirth purohits* and the Samiti surrounding the question of who, under what conditions, could facilitate access to the inner sanctum of the temple by allowing people to enter through the side door of the temple and not wait in the queue. *Tirth purohits* wanted to be able to take their important patrons through the side door. I remember that what struck me about this scene was not the confrontations themselves. After all, there are many contexts when matters having to do with long lines and short tempers become confrontational. What I found notable was that, in the half circle of onlookers that surrounded the small nodes of tension there were many pairs of men holding hands as friends do, and each pair consisted of one *tirth purohit* and one employee of the Samiti. To me, it was an affirmation of the solidarity of local identity (in both groups almost everyone hailed from the Kedarnath valley) against bigger and in many ways translocal pressures of the high season that were pitting them against each other with so much force. The pressures of the business of *yatra* were changing Kedarnath, and tension-filled moments like this were becoming more common.

LABOR

Kedarnath in 2007 and 2008 was a place that highlighted the ability of pilgrimage tourism to radically change the economic, cultural, and environmental landscape. In addition to the pilgrimage priests and Samiti employers there were all the people who worked in shops and lodges. A community of cleaners came on contract from Uttar Pradesh each year and lived in Kedarnath. There were employees of the Garhwal Regional Development Authority (Hindi: Garhwal Mandal Vikas Nigam or GMVN) and doctors and nurses who worked in the private clinic and the small public hospital. There were the employees of the helicopter companies. There were the dozens of small chai shops and thousands of rooms for rent between Guptkashi and Kedarnath. There were the employees of the Public Works Department who maintained the road. If one extended the unit of analysis to include the entire Uttarakhand Char Dham and the Sikh pilgrimage destination of Hemkund Sahib, then the economic zone affected by Char Dham Yatra tourism would also have to include the entire national highway and rail network between Delhi and Haridwar traveled by most of those on their way to Uttarakhand. The entirety of this built and economic environment rested on the shoulders of those involved with transportation and construction. With the exception of locally grown fruit, vegetables, rice, dal, and wool products, almost everything had to come up into the mountains by truck to somewhere between Guptkashi and Gaurikund. Thence, depending on its final destination, it was carried the rest of the way by men, ponies, and mules from typically either Sonprayag or Gaurikund. The only exceptions to this rule were materials that arrived by cargo helicopter, a rare occurrence that became increasingly common after 2014 with the construction of a new helipad behind Kedarnath that was able take the weight of the cargo helicopters. The inadequacy of the old one was greatly apparent during the rescue and relief operations of 2013, when the waterlogged ground could not take helicopters' weight.

A staggering amount of labor was involved with the delivery of materials to Kedarnath. Often visitors would be aghast at the high prices charged for consumable goods in Kedarnath. While there was a certain degree of price inflation that had to do with a fairly ideal seller's market, much of the inflation was based on how much it cost to get the goods to Kedarnath in the first place. Men, mules, and ponies were also many *yatris'* means of journeying from Gaurikund to Kedarnath. Small adults and children would often be carried on the back of a single porter, and teams of four men would often carry *yatris* in a chair slung between two poles that in form greatly resembled the *dolis* of the *devtas*. Many of the porters were Nepalis who annually came to Uttarakhand to work during the pilgrimage season. Most of the horse and pony drivers were Kedarnath valley locals. Porters traveling back and forth from Gaurikund to Kedarnath would often make several trips a day, a level of physical exertion that I found difficult to comprehend.

There was also, always, a great deal of construction going all along the motor road in the Kedarnath valley and in Kedarnath as well. The years 2007 and 2008 were a boom time for construction. People were building new rooms and lodges, renovating existing buildings, and adding amenities that would be attractive to middle-class *yatris*. Friends of mine were deciding against moving down to a city like Dehra Dun and looking for work and instead borrowing money to build a lodge with attached restaurant in a good spot along the Guptkashi-Gaurikund road.

LIVELIHOOD

Kedarnath had become a place for big business by 2007. This was an important aspect of the feel of the place—the constraint and tension produced by economic forces. The following is a close, remembered paraphrase of what a local Kedarnath valley man told me when we were sitting conversing and having chai together in Kedarnath:

> Whoever lives in Kedarnath, his first thought is for his livelihood. First livelihood, then Kedarnath. When I was little I used to see [the renunciant] Phalahari Baba, and since then I've never seen any other saint like him. His burial place [Hindi: *samadhi*] is on the way to the Bhukund Bhairavnath path. He used to live here twelve months out of the year. His sitting place was next to a shop across from the temple, and I along with two or three other boys would go and steal ten-rupee notes from underneath his knees. We used to do this three or four times a day, always in front of others who were sitting with him. Seeing us, they would start to scold because they knew we were coming to steal. Baba used to smile and scold the scolders, saying, "Why are you scolding them and preventing them? What have they done to you? Let them come."
>
> Nowadays whoever is living here, everyone's first priority is money and then Kedarnath. No one can say that I am here for Kedarnath. If anyone says this, they are out-and-out lying. There was once a shepherd who used to live near Vasuki Tal, originally from Himachal. So once I went to Vasuki Tal and drank chai in his hut and he gave me two roti to eat. I became friends with him. And whenever he came down he would always come visit me in Kedarnath at least once for chai and conversation. One day he told me, "I saw an actual [Hindi: *sakshat*] Kedarnath temple near Vasuki Tal. *Puja* was happening there exactly in the way that it happens here in Kedarnath." He told me but I didn't believe him. But he was totally certain, with full belief [Hindi: *vishvas*], that he had seen what he had seen. But one second later nothing was to be seen—it was just a glimpse. So I said, "You have had a vision of Future Kedar [Hindi: *Bhavishya Kedar*]. In the *Kedarakhanda* it is written that here there is somewhere a Future Kedar, just as there is a future Badrinath [Hindi: *Bhavishya Badri*]." When I was once near Vasuki Tal I was in my own devotion singing a *bhajan* with a soft voice and then with increasing volume. Then a *devta* came upon a woman who was sitting near there. She started screaming loudly. I saw this, so I thought I should stop singing that, so I stopped.
>
> For *yatris* this place is heaven, but for us it is a place to earn a living. Also *sadhus* are coming here for money. If I did not need money, I would come to Kedarnath for a week or so and then leave. It is of course true that there is a great deal of faith [Hindi: *shraddha*] in Kedarnath because everyone depends on Kedarnath for earning, but where there is livelihood there cannot be [religious] feeling [Hindi: *bhav*]. When I am thinking about earning it is impossible to relate to Shiva appropriately. While it is true that during Shravan I will go pour water on the *linga*, I basically do this only because I am there already. I greatly esteem Kedar [Hindi: *Kedar ko bahut manta hun*], but most of the time when I am here I am thinking about *yatris* and not my own *bhav*.

This local man agreed with my perception that all of the development had happened in the last ten to twenty years and especially in the last five or six, but he did not see the development as being connected to the creation of the state.

He said he had formerly wandered a lot but now could not go up very far from Kedarnath itself. There was something special in Kedarnath, and whenever one left Kedarnath even to go as far as Garud Chatti this peace (Hindi: *shanti*) was lost.[18]

TREKKERS

One of the most important ways in which Kedarnath, and indeed the entire landscape of pilgrimage tourism in Uttarakhand in recent decades, had been changing was that it was seeing a marked increase in the number of trekkers and visitors whose primary goal was to walk in the mountains and to raft the rivers. In Kedarnath such visitors were sometimes interested in a half-day jaunt up to Gandhi Sarovar and sometimes were experienced mountaineers keen to walk on a glacier. Most of them used to come up the path from Gaurikund with everyone else. However, on June 14, 2007, three Bengali men and six porters arrived in Kedarnath by foot from the north, along with one renunciant. They had walked over the mountains from Rhansi in the Madmaheshwar valley, two valleys to the east. They said that it had taken them nine days and that they had walked on a glacier for three or four kilometers of the trip and had narrowly escaped death several times. At one point, just after having performed a *puja* at a local goddess shrine, they heard the sound of a breaking glacier on the other side of the mountain from where they were. Camping by tent, they said that they had come to Kedarnath via the Great Path. On another occasion, a group entered Kedarnath from the north and reported that they were returning from a month-long journey to Kailash Mansarovar. Numerous sightseers would make the short trek up to Gandhi Sarovar or go walking in the side-valleys behind Bhukund Bhairavnath. The occasional parade of groups of trekkers into Kedarnath was part of the *mela*— the difference was that in some cases they explicitly were not in Kedarnath out of *bhakti* or out of desire to take *darshan*.

GARHWALI SUPERNATURAL BEINGS

What non-Garhwalis might see in a relatively high Himalayan landscape differs radically from how such locales are imagined in traditional Garhwali worldviews. The end-valley of Kedarnath is significant in Garhwali religious worldviews for more than the journey of the Pandavas and the presence of Shiva, Devi, and Bhairavnath, as I found out during a monsoon-season chat in Kedarnath. On July 22, 2007, in the month of Ashadh, when Kedarnath was almost at its emptiest and there were many rainy days of chai shop conversation, Bhupendra and I found ourselves in a conversation with the elderly mother of a Kedarnath *tirth purohit* friend and two of her female friends at his *dharamshala*. She had been in Kedarnath for a month and planned to stay until the end of Kartik.

FIGURE 12. A Brahma lotus.

During the conversation I asked her if Kedarnath and Madmaheshwar were equally important, and she said, as a way of answering, that if women from the Madmaheshwar valley did not sing *payari geet,* a Garhwali genre of auspicious (Hindi/Garhwali: *mangal*) songs that greet the deity when it travels to the village, then the god cursed them but that she (as someone from the Kedarnath side) usually sang Garhwali *bhajans* about Shiva. She knew *payari geet* for Madmaheshwar but did not want to sing them because we were not in Madmaheshwar's territory.[19] Garhwali deities are notoriously territorial. She also said that if she sang them in Kedarnath the *acheri* would come for her. *Acheri* is a Garhwali term for a class of forest-mountain spirits. In the high places and wooded river banks of Garhwal there is often the possibility that someone will attract the attention of such spirits. For example, the life story of one of the folk heroes of Garhwali literature, Jeetu Bagadval, ends with his being kidnapped by *acheris* because his lovely flute playing attracted their attention. In the Kedarnath valley it is customary to keep quiet and not wear bright colors when traveling at high altitude for this very reason. She said that once she had gone with a group of women above the Bhairavnath shrine to gather Brahma lotuses (Hindi: *brahmakamal*). Brahma lotuses grow only above about 4,200 meters and are offered, either fresh or dried, in the Kedarnath temple. When you pick them you must go barefoot, after having bathed and without having eaten. It is a special form of place-specific *prasad* (Pinkney 2013b). Pilgrimage priests both give the lotuses to their patrons and sell them to others. Everyone was keeping quiet lest they be kidnapped by *acheri.* Then near a river someone called her name a little loudly and she became sleepy and started to feel strange; she thought that perhaps she was about to be kidnapped. But they came back down near Bhairavnath and called on him for protection and she was fine again.

NATURE AND *SHAKTI* IN KEDARNATH

Sometimes, in conversation with both Kedarnath residents and visitors, I would hear a clear distinction made between the place's natural beauty (Hindi: *prakritik saundarya*) and the power transmitted into the place by Shiva's presence. More common, however, was a complicated explanation that involved the overlap of these two modalities of place-engagement. Here is an example of what I mean.

During the monsoon season the numbers of visitors from the North Indian plains declined markedly and the hectic rhythms of the high season slowed. It became easier to have longer conversations. I had one such conversation with an older man from Bangalore who made his living selling flowers for use in *puja.* I met him around the time of the Samiti-sponsored recitation of the *Shiva Purana* on August 22, 2007, just before the *Bhatuj* festival. I spoke with him in the presence of the *tirth purohit* owner of the *dharamshala.* Over the years the two had become good friends. The Bangalore *yatri* said that he had traveled all over India and Nepal and had been coming to Kedarnath since 1983, when it had featured minimal *vikas* (Hindi: development) and very few people. According to him, most of the changes had happened in the last seven or eight years. Every day when he visited Kedarnath he would walk out behind the temple to the river banks behind Kedarnath village to sit, perform meditation (Hindi: *dhyan karna*), and occasionally sing devotional songs (Hindi: *bhajan*) to nature and/or deities, a conflation that he himself made. He had started staying in Kedarnath because he felt a special attraction and because he felt peace (Hindi: *shanti*). He gave an amazing definition of *shanti,* explaining that there are two things in Kedarnath—a "magnet power" that attracted people and natural beauty. When the thread (Hindi: *tar*) of those two things became one with the *tar* that is inside a person, then the feeling of *shanti* was produced.

Not everyone experienced this kind of powerful conflation of divine power and natural environment. Some people attested to the opposite. On August 25, 2007, I spoke with a Virashaiva renunciant who had been staying in Kedarnath for some time. He did not find stories about the Pandavas or the practice of *ghee malish* to be particularly important. He felt positively affected by the natural beauty of the place and said that, as in Rameshvaram (which looks out onto the Indian Ocean on the other end of the subcontinent in the south), he felt as if he were on the edge of heaven. He emphasized that for him the actual Kedarnath *linga* was not the center of his religious attention because like all Virashaivas he believed that Shiva was actually formless and *lingas* were simply meditative aids. He directed most of his devotion (Hindi: *bhakti*) toward the small *linga* that he wore around his neck (called *ishtalinga* by Virashaivas) and had been given by his guru. He performed *puja* of this *linga* daily, and when he performed *puja* and meditated it shone for him as he held it in his hand. Kedarnath was of special importance for Virashaivas, he said, because the first guru of the Virashaivas, Ekoramaradhya, first manifested into the world by emerging out of the Kedarnath *linga.* Another renunciant who spent the entire season in Kedarnath, performing at least 150,000 recitations of the *Om Nama Shivaya* mantra per day, expressed a similar sentiment to me. He said that he entered the temple more than most *sadhus* but that even he did not feel a strong necessity to go inside because he was already living in the center of Shiva-terrain (Hindi: *Shiv-bhumi*).

I happened on another hyperarticulate explanation of the distinct power and character of Kedarnath in late September when I met with a group on a visit to Kedarnath with their guru. The guru had come with his followers for a five-day stay. His followers consisted of singles, couples, and families from both India and America who ranged in age from approximately the midtwenties to the midsixties. They picked several different spots (near the helipad, by Bhairavnath, Shankara-charya Samadhi, Phalahari Baba Samadhi) for morning-long sessions of *bhajan* singing, teaching sermons (Hindi: *pravacan*), and question-and-answer periods.

I visited two of these sessions. Most of the group were dressed in trekking gear. During the first session I saw an example of what I would term *guru bhakti,* or *bhakti* directed toward the *guru.* An elderly man sitting near the guru stretched out his arms and, with slants and twists of his torso and movements of his hands, began to tilt into a horizontal position that gradually brought him closer and closer to the guru. Eventually he clung to his *guru,* shaking, sometimes resting with his head in his lap in great contentment and sometimes hugging him with desperate force. At one point it looked to me as if his joy was so total that he lost bodily control. The guru remained calm throughout, and it appeared to be for him, as for most of the group, a fairly normal occurrence.

The second session I had with the group was behind the Bhukund Bhairavnath shrine. When I arrived around ten in the morning a *pravacan* was in full swing, mostly in English. The guru was telling the story of Daksha and Sati with an eye toward the moral that one should always respect the saints and deities of all religions and that if you do not then bad things happen. He described some of the other holy places in the Kedarnath area and the special powers and vibrations associated with them. He dwelt on the importance of Mount Kailasa and narrated at length his own spiritual journey, emphasizing that it had been accomplished through *bhakti* and *bhajans* and not through yoga and meditation. He was quite clear that those techniques were not his path and that instead it had been through *bhakti* that Shiva had at times awakened all seven of his chakras in an instant. He described various visions and levels of experience such as the contemplation of a thousand-petaled lotus of light in which each petal is a different color shaped as a triangle, at the center of which would be the deity of one's choice.

He spoke of the necessity to spend as much time and effort striving for God as one does striving for anything else, and said one must be ready to stand before God and answer the question "What have you done for me lately?" Toward the end of his talk he began to answer questions, and the question of language arose. He said that Hindi was the best language for talking about religion, especially the pure Hindi that was descended from Sanskrit, though if he spoke that way no one would understand. He said that Punjabi was not very good in this regard but better than English. He gave the example of the word *joy.* In English what does *joy* mean? Nothing. But *anand* (Hindi: joy, bliss)—that word has a range of important meanings (such as, for example, the idea that reality as a whole is fundamentally made of joy). He then began to sing devotional songs, beginning with an Urdu *ghazal* that I did not understand, then shifting to *Om Namah Shivaya,* and finally to *Raghupati Raghav,* a popular *bhajan* associated with Mahatma Gandhi that is often popular for its ecumenical message. He then closed his eyes for over ten minutes, during which time tears began to flow from his eyes. Some of his followers followed his example and remained steadfast in their own practice, while others started to put on warmer clothing because fog had started to come up the valley and the temperature was dropping.

When he came out of his trance and began to walk back down into the village I introduced myself in Hindi. He smiled and patted my shoulder and said he had already heard of me and what did I want to ask him. My question was interrupted by a renunciant who with choked-up voice and propitiatory tone began a tragic story involving a girl, marriage, and money. The guru responded sternly, saying, "My giving you money will not help you." Then he said, "I am Shiva, I am a destroyer and I have come to destroy. I am not Vishnu, I do not take care of people [Hindi: *mai phalan nahin karta*]." The *sadhu* become even more emotional, calling the guru Lord (Hindi: *Prabhu*) and saying that he could not hide anything from him. Several devotees exchanged knowing smiles. Then I asked the guru what he felt was special (Hindi: *vishesh*) about Kedarnath and how he understood the connection between the land and the god. He replied that God and *shakti* are available everywhere but the difference is that in Kedarnath (as in three other places: Dwarka, Rameshvaram, and Vishvanath in Varanasi) the *shakti* is awake (Hindi: *jagrut*). In Kedarnath looking for *shakti* is like digging for water in a place where water runs close to the surface. I asked him what he felt about the ritual of *puja*. He said that rituals such as *puja* are not intrinsically important; it is the feeling (Hindi: *bhav*) that is important. This conversation lasted from the shrine of Bhairavnath until we entered the bazaar. His devotees were crowding around him and one was kissing his shoulder. As we arrived at one of the restaurants in the market it was clear that our time was over.

FORESHOCKS

Looking back on the 2007 season in the aftermath of the 2013 floods, I feel as if there were a series of premonitions, or, to apply a geological term to a set of socio-ecological phenomena, foreshocks that (in retrospect) were leading up to the floods of 2013. I have already sketched something of the social, economic, and ritual tensions generated by the high season. The end of the season saw additional anxieties. Near that time an official from the state forest department arrived in Kedarnath and began to loudly proclaim that Brahma lotuses were being overharvested. As the official strode through the bazaar, Kedarnath locals began to quietly pull out their phones. It was revealed that he was not there in any official capacity. He was trying to line his pockets while he was still able to say that he worked for the state. His aims unveiled, he hurried out of Kedarnath. At the time I was, perhaps naively, a bit confused. No one seemed concerned at all that perhaps people were in fact picking too many Brahma lotuses. Later that day, sitting in his shop, a Kedarnath pilgrimage priest told me that no one was willing to discuss such things in public because privately everyone was worried that the Mandakini would dry up and the place would become uninhabitable. Such concerns were publicly unvoiceable.

On October 31, just before the end of the season, one of the long-term resident *sadhus* in Kedarnath who also spent part of the off-season in and around Ukhimath came into the market and started going on loudly about how a ghost was bothering him: that of a local youth who had died in a freak accident while assisting passengers in getting out of the helicopter earlier in the season. He said that the youth "keeps coming to my tent and trying to talk to me, but he speaks in Garhwali and I can't understand him because I'm not Garhwali. I'm leaving early this year." This *sadhu*, who had been coming to Kedarnath for years, said that he normally stayed until the last day of the season in order to, among other reasons, receive gifts from wealthy *yatris* (he referred to them as *yajmans*, patrons) who came annually for the closing *puja* and ceremonies.

This ghost incident was apparently part of a larger set of occurrences in which the dead youth had been recently pestering people. One man told me that he was often touched, jabbed, and poked at night by ghosts, though another said he walked at night where he wished and did not fear ghosts in the slightest. As it got colder and the season began to wind down, many locals quietly went home a bit earlier than they had in previous years; the presence of the dead youth was one of the reasons cited by those commenting on the early departure of others. As the last resident in the pilgrim rest house where I was staying, I was repeatedly asked whether I was frightened to stay there alone at night. Eventually I did become frightened one night. Other locals told me that they had not seen any ghosts in Kedarnath and felt no fear at all when walking around alone at night or in their rooms.

This occurrence activated a set of memories and tales regarding the Kedarnath of previous decades. Many pilgrimage priests and other locals could recall a time, beginning several decades earlier, when the village of Kedarnath had had far fewer buildings, lodges, electricity, and lights, when latrines had all been outside the boundaries of the village, and when residents had been afraid to venture beyond the lit boundaries of the village for fear of ghosts (Hindi: *bhut*) and other supernatural beings such as *acheri*. Even rumors of the appearance of this dead youth made this older Kedarnath seem less in the past, more in the present. It occasioned an active nostalgia of sorts. Yet at the same time it was the modernized, present-day face of Kedarnath in the form of helicopter blades that had created the situation in the first place. The death had not been the result of a flood, a landslide, an earthquake, a bus accident, a fall due to drunkenness, or a lack of timely medical expertise—these were all known possibilities. It was something new, a deadly sign of how the region was changing.

In Garhwal the deceased usually become present and able to communicate through possession, but this was not a standard case. D. R. Purohit, noted scholar of Garhwali folk-life and religion, confirmed my assessment that this case was, in Garhwali eyes, nonstandard. He informed me that normally, in such cases, the dead youth would have first signaled his presence by causing someone to weep violently. This would then have required that the ghost be caused to dance (Hindi: *bhut nacana*). Through this procedure, presided over by a *ghadyalya* (a particular type of Garhwali ritual specialist) and conducted in the home, the young man would have possessed some individual from his own home (not from Kedarnath, but from his own house) and spoken through him or her. The ghost would then have lamented his own fate and the fate of his dependents (if any) and relatives and would have stated his unfulfilled wishes. This procedure would have been performed every third or fourth year until the dead person was appeased, at which point an image of the ghost would be made and immersed in the marshes and water tanks of Kedarnath.[20]

Thus the way that the dead youth made his presence known was not the way in which he might have been expected to make his presence known. He spoke without being invited to speak. And he appeared in a place unequipped, in a Garhwali sense, to respond to his attempts at communication. In retrospect, this whole occurrence seemed like a premonition of how the pull of Kedarnath was transforming the Kedarnath valley.

ECO-SOCIAL ENTANGLEMENTS

Looking across these vignettes, one can see a diverse set of vectors—human, nonhuman, material, nonmaterial, organic, inorganic, animate, inanimate, cultural, ecological, divine, devotional, human, economic, natural—that are, recalling Edward Casey's notion of place as that which *gathers,* pulled together in a set of complex entanglements of increasing intensity. No one factor of human engagement with and experience of the place can be wholly disentangled from others, just as the changing place cannot be disentangled from the human activities involved in the always-emergent changes. In these vignettes, what is inside the temple cannot be wholly separated from, or wholly conflated with, what is outside the temple. In these scenes the effect of the weather overlaps with the effect of economic constraint, both of which add to the *shakti* of the place/*shakti* of Shiva in the place, a *shakti* engaged and imaged through *puja, darshan,* meditation, singing, bathing, and wandering. This was the Kedarnath I saw in 2007, 2008, and 2011.

5

When the Floods Came

Water, rock, and mud descended on Kedarnath twice, first in the evening of June 16, 2013, and then in the morning around seven on June 17, 2013. Those hours are in fundamental ways beyond discursive reach. But we can try to gather fragments, as have dozens of video documentaries, hundreds of journalist accounts, and thousands of Facebook posts. In a fragmentary way, it is possible to say that Kedarnath remained, during and after the flooding, a place that resisted human control and a place marked by a relatively high level of shared experience across diverse social, economic, and cultural subjectivities. It remained a zone of transition to the beyond-human, a vanishing point relative to human frameworks of understanding and control. It remained a place characterized by the enmeshing of the divine and the natural. The floods came as a surprise and a rupture, but also, given their Himalayan location, they were not a surprise and they were not a rupture.

I embark on the description of this event with the concept of "border situation" in mind as it is discussed by anthropologist Michael Jackson in his work *The Palm at the End of the Mind*. Jackson (2009, xii), as part of a broader engagement with the question of the putative cultural universality of something called "religious experience," found himself focusing on "critical situations in life where we come up against the limits of language, the limits of our strength, the limits of our knowledge, yet are sometimes thrown open to new ways of understanding our being-in-the-world, news ways of connecting with others." He was drawn to the term *situation* and its use in compounds like *limit situation* (from Karl Jaspers) and *frontier situation* (from Theodor Adorno) because the concept of situation is an "existential" concept that engages experiences of being-in-the-world in a way that does not reduce those experiences to either "political economy" or "religious belief or doctrine" (Jackson 2009, xii). I am not focused in this work on arguing for or against the specialness of religious experience. However, being in Kedarnath in the best of times has constituted, I think, the beginnings of a border situation for many in which a combination of multiple elements (some of which are understood to be "divine") of the situation exert ontological weight. And this, awfully, would have been even more the case during the floods. In this chapter, I treat the first days and weeks of the disaster (Hindi: *apda*) as a sort of border situation. My primary goal in this chapter is not to critically assess the relief and rescue measures undertaken by the state government, the national government, and other groups, though a critique is absolutely warranted. Nor is it to offer, as Michelle Gamburd (2012, 2) does regarding the impact of the 2004 tsunami in Sri Lanka, the first stage of an argument about the "social consequences of natural hazards and the cultural processes at work in humanitarian relief efforts" and to analyze closely the nature of relief efforts and funding in the aftermath of the tsunami. My primary goal is to attend to the feeling of the situation.

Among scholars of disaster, it is conventional wisdom to note that, while disasters are often experienced and pictured as catastrophic and surprising events of short temporal duration, close analysis often reveals that the conditions that came together to produce the experience of what is often called a "natural disaster" sprang from a deeper and broader set of causes such as (often willful) human ignorance or selfish, unjust myopia about safe, long-term strategies of residence and anthropogenic landscape change in a specific location, and that consequently the disaster might be better termed an *unnatural* disaster, or a disaster with an "unnatural history" (Steinberg 2000). Such events are properly understood in the "*longue durée*."[1] Anthony Oliver-Smith (1999a, 75), for example, in his analysis of a terribly destructive earthquake in Peru, wrote that "the destruction and misery in Peru in 1970 and after were as much a product of that nation's historic underdevelopment as they were of the earthquake" and that this underdevelopment was tied to Peru's colonial past. The human vulnerability that is on terrible display during these events is, in many ways, a vulnerability that has arisen out of a specific set of longer-term eco-social conditions. But this way of understanding disasters does not mean that we should not attend to the events themselves and how they were experienced at the time. Therefore, in this chapter (complemented by the longer-term analysis offered in the next chapter) my goal is to attempt (and of course necessarily fail) to give a sense, from a distance, of how the *places* of Kedarnath and the Kedarnath valley may have "felt" at this time and to contextualize what they felt like at this time relative to the broader life of the place in a way that captures something that people who survived would recognize. I was not in Kedarnath in June of 2013, though had travel plans turned out differently I might have been. I therefore drew on media accounts, phone and e-mail conversations, eyewitness accounts on Facebook, and government reports in assembling what follows. In the next chapter the analytic view again widens.

Imagine that you were a visitor to Kedarnath. You were at approximately 3,500 meters in the Himalaya and a cold rain had been falling continuously for days. You had just completed a demanding journey. You probably had come in a small bus or car from somewhere between Rudraprayag or Guptkashi the day before, passing most of the day (or night) on a narrow mountain road with a great deal of traffic. The day before that, you had come south from Rishikesh or Srinagar or east from near Uttarkashi. The road had taken you only as far as Gaurikund. The mostly paved path from Gaurikund had been muddy and slippery, the rain partially masking the smell of horse and pony urine. If you rode, as the majority of visitors had begun to do in recent years, then your thighs hurt and there were many moments of anxiety that the creature would fall into the ravine and thence into the Mandakini River. Many people could not mount or dismount from the horse without assistance. If you were a woman this might have meant that you needed to accept the somewhat intimate help of a strange man. If you came on foot then you were mentally tired from avoiding being pushed over by ponies or visitors being carried in palanquins. You would have ascended over 1,500 meters in the last day or two. You might feel sleepy or dizzy or struck with headache. You might be so overcome at finding yourself in Kedarnath and so focused on reaching the temple that you were unaware of any of the environmental factors described in this paragraph. If you flew over all of this and arrived by helicopter it meant that you did not know when you would be able to leave—it was raining so hard that helicopter service had been stopped.

You found yourself amid a press of people because it was the high season. Approximately ten thousand visitors and hundreds of temporary residents were together in an area of approximately two American city blocks square. Many rooms were going for over one thousand rupees per night, and it was difficult to find a place to stay without prearrangement. You were surrounded by families with schoolchildren on vacation. If you were in Kedarnath or the upper Kedarnath valley and you were not a visitor, then this time would have been busy, challenging, and draining. For residents of the Kedarnath valley and other Uttarakhandis who made their living from the pilgrimage tourism industry, as well as porters (often from Nepal or Uttar Pradesh) and sweepers (from Uttar Pradesh) who came to the Kedarnath corridor solely for employment, this month and a half provided most of the income for the year. That meant that people had been constantly on the go for over a month in a high-altitude, high-stress environment with a great deal of money at stake.

Social tension was part of the weather in Kedarnath in mid-June of 2013. The porters were on strike, protesting the growing amount of helicopter service to Kedarnath and the special privileges accorded its passengers. In mid-June of 2013 this meant that many visitors to Kedarnath found their existing plans suddenly interrupted by the unavailability of horses and ponies. If you were only just arriving in Kedarnath at some point in the morning then quite possibly the line would have been too long for you to take *darshan* in the inner sanctum on that day. So Kedarnath was exceptionally full: of people, of *shakti,* of rain, of tension. Many local youths were also visiting Kedarnath either for pleasure or for short-term work because they had just finished their semester at intercollege.

THE FLOOD EVENTS

It had been raining very heavily since the day before. The monsoon came to the area several days early. Monsoon rains crossed Uttar Pradesh and were in the mountains in a single day. Monsoon weather coming from the Arabian Sea in the west and the Bay of Bengal in the east suddenly bracketed Uttarakhand. In a very short period of time, Kedarnath arguably received "more than half the rainfall Delhi receives in an entire year" (Mihir and Irani 2013). Behind Kedarnath by this time the rains had already caused a great deal of river gulley erosion. Accumulated sediment was making it difficult for the extra water to flow naturally, and instead it was building up (Dobhal et al. 2013, 173–74).

It is necessary to understand how Kedarnath is situated geologically. It is located on the "outwash plain" of several glaciers that lie to the immediate north of the site (Dobhal et al. 2013, 173) and rests on somewhat unstable layers of rock, dirt, and mud left over from previous glacial movements. The Mandakini and Saraswati Rivers flow down through glacial moraine after emerging from the glaciers. The main course of the Mandakini flows to the west of Kedarnath. The main course of the Saraswati used to flow to the east of Kedarnath but in recent years had been diverted by barricades and was now flowing southwest into the Mandakini, joining it just above and behind Kedarnath. That meant that directly behind and rising above Kedarnath at steep pitch were small mountains of ice and rock out of which two rivers were flowing. Each of these rivers had a small network of smaller gullies that would not fill with water unless the river was in spate. This entire area had been filling up with water all day because of the massive rainfall.

As the situation unfolded, it quickly became apparent that Kedarnath, and Uttarakhand more generally, were undergoing extreme flooding. I will use part of an account written by Pyoosh Rautela (2013, 45), director of Uttarakhand's Disaster Mitigation and Management Centre, to structure the discussion of the flooding and its immediate aftermath:

> Most evidences on ground having been obliterated by flood it is important to look for circumstantial evidences in the narration of the survivors of the deluge so as to reconstruct the sequence of events. As narrated by the survivors of the disaster;
> i) there were incessant rains in the area between 14 and 17 June, 2013, ii) rainfall

on 16 and 17 June, 2013 was particularly heavy, iii) tragedy struck Kedarnath on the night of 16 June, 2013 and in the morning hours of 17 June, 2013, iv) hitherto abandoned eastern channel of Mandakini at Kedarnath became active in the evening hours of 16 June, 2013, v) flooding in Kedarnath was not that devastating on 16 June, 2013 though it washed off the pedestrian bridges over Mandakini connecting Kedarnath to Rambara and turning Kedarnath into an Island, vi) on 16 June, 2013 flood waters washed off Sanatan Dharm Sabha guest house, Shankaracharya Samadhi and a few other structures in the vicinity of the temple together with a few persons, vii) floodwaters of Mandakini did not affect the Kedarnath temple premises on 16 June, 2013, viii) after the flood event, despite heavy rains, most persons in Kedarnath assembled in the temple premises and engaged in prayers.

At around 5:15 p.m. things started to come to a head (Dobhal et al. 2013, 173). The volume of rain and glacial runoff caused these existing gullies to erode. Debris from the gulley sides, combined with the presence of the barricade, meant that much of this water began to pool above Kedarnath rather than flow down the valley. When this pool built to a sufficient volume it surged to Kedarnath and "simultaneously picked [up a] huge amount of loose sediment en route" (Dobhal et al. 2013, 174). This surge of water and debris washed away the structures in the northwestern part of Kedarnath and, much of its force undiminished, continued downstream. Different eyewitness accounts place this surge within a two-hour window, but most agree that it happened while *shringar arati* was occurring. The temple became a gathering place for much of the night. Many people stayed there until early morning for safety, comfort, and prayer. A renunciant who survived the floods told me that he spent that night in his customary location just to the east of the temple listening to the screams and cries of those in the temple, a memory that (combined with the trauma of surviving the two flood events) drove him out of his mind for months. There were now two rivers flowing in spate, one on either side of Kedarnath. And the bridge across the Mandakini connecting Kedarnath to the path down to Gaurikund was gone. This surge of mud and water continued downstream and erased the village of Rambara.

I came across a literary description of this time in Rambara in a book entitled *Apda Ka Kafan* (Disaster's Shroud; Pāṇḍey 2014, 4–5). The title of the section was "Trahi Mam." This phrase is part of the mantra, discussed in chapter 4, that is recited along with the *ghee* massage in the inner sanctum of the Kedarnath temple:

> I am a wicked person [Sanskrit: *papa*], I am one whose actions generate wickedness, my inner being [Sanskrit: *atma*] is full of wickedness, I am one from whom wickedness arises. Save me [Sansrkti: *trahi mam*], lord of Parvati! O Shiva, you destroy all *papa*.

The few who survived the situation in Rambara managed to quickly climb the steep sides of the valley amid rushing water, mud, and landslides. The majority of those in Rambara did not survive. The lower half of the river valley between Kedarnath and Gaurikund, and continuing down to just north of Sonprayag, is quite deep and narrow. This would have focused the force of the surge as it rounded the corner into Gaurikund. It tore away three-quarters of Gaurikund and roared into the flat and open bazaar of Sonprayag, which had been built that way to accommodate the hundreds of buses, cars, and jeeps that pass through during the high season, and swept vehicles and houses on the east (river) side of the village into the churning waters. It weakened the hillside and caused landslides all the way down the Kedar valley. It submerged a recent hydroelectric project built at Kund, where the road forks, with one road leading to Guptkashi and the other road to Ukhimath and, via Chopta, Badrinath.

When I write these things about Rambara, people and scenes flash in front of my mind's eye. Many of the shopkeepers in Rambara were from the Ukhimath side of the valley, and I felt a personal connection with them. In my own foreign way, I am more connected to the Ukhimath side because that is where I lived during the off-season. I once spent three hours warming up in a cigarette shop no bigger than the width of two people because the proprietor had seen me many several times in his village of Kimana, one of the villages that make up the village cluster known as Ukhimath. Bhupendra's father owned a shop in Rambara that he rented out. Bhupendra, with whom I lived and worked in Kedarnath for a season, was also from Kimana, so this brings me to what happened to him. The first flood event took him in his room on the northwestern side of Kedarnath, I was told later. He was working for one of the private helicopter companies in Kedarnath and was off duty at the time. I suspect he would have been cooking dinner, as he taught me to do. I am sure that if he had time he tried to save others around him. Also lost at this time were Mahant Chandragiri and his *langar*, and Bharat Sevashram Sangha. Many *sadhus* built temporary shelters along the path that went from near the free food kitchen northeast down to the temple. During high season the queue for *darshan* would stretch around the corner, and the *sadhus* would be able to interact with and receive donations from *yatris* who were waiting in line. Most of these sitting Shivas would not have had time to elude the waters.

Early in the morning of June 17, most people thought that they were past the worst and that the worst had been a recognizable disastrous event: flooding.

x) Major devastation took place in Kedarnath in the morning hours of 17 June, 2013, xi) Chorabari Tal was intact on 16 June, 2013, xii) Rambara and Gaurikund were devastated in the night of 16 June, 2013, xiii) breach of Chorabari Tal took place around 0700 hrs on 17 June, 2013 and xiv) floodwaters of Mandakini ravaged Rambara, Gaurikund and Sonprayag again in the morning hours of 17 June, 2013. (P. Rautela 2013, 45)

But what happened at around seven that morning was something different. Intense rain had continued to fall, snowmelt had continued, and the volume of water in the lake known locally as Chorabari Tal (and more famously as Gandhi Sarovar) continued to rise. Chorabari Tal is a famous destination for those who arrive in Kedarnath with energy to spare who wish to take in the "natural beauty" of the Himalaya. It is a lake that is filled by glacial runoff and may be mostly empty or quite full depending on the time of year. Normally, excess water in the lake would have seeped out and down through the network of holes and gullies in the moraine barrier that constituted a major portion of the boundary of the lake and part of the debris layer on top of the snout of the Chorabari glacier. But all of the rain and previous erosion and accumulation of debris meant that there was no place for the massively accumulating volume of water to go. It broke. "_____" was the sound it made. A wave of water, mud, rock, and debris rushed down from the lake to which I had once walked with Bhupendra (whom people often called Bhupi for short) and where we had once written "Luke and Bhupi" on a rock. Many people have described the wave as more than ten feet tall. Two scientists from the Wadia Institute of Himalayan Geology had been at their monitoring station near Chorabari Tal and saw it happen.[2]

People had a split second to make a decision about what to do. Many who survived scrambled up the hillside toward the Bhukund Bhairavnath shrine. Others found a roof that, for the fortunate, did not later collapse. Some took refuge in the temple. When I visited Kedarnath and the Kedar valley I heard many accounts in which people had saved someone (a child, a relative, a stranger) from the flowing mud-rock but had been swept away themselves. A Guptkashi man in Kedarnath at the time could say only that it was "beyond imagination" (Hindi: kalpana ke bahar). When I see people talk about it in person or in television interviews I see often see the same facial moment—a quick tightening of the face, a ripple of intense emotion, and then diminished affect.

I know what I know about this event in different ways. I spoke with some people on the phone in the weeks following the floods. I read accounts online and watched interviews on YouTube, often with people whom I recognized from my fieldwork. When I managed to speak with survivors in person for the first time a year after the event, I often could not bring myself to ask people to narrate what they experienced. I did not want to ask them to go back to that moment. I remember what a friend of mine said to me after he told me on the phone that he had been in Kedarnath and survived. He said, "I saw Shiva's tandava." I said, "I cannot imagine it." His response was something to the effect of "Good. That is right. You can't." Now, as I write, I cannot remember the exact words. I did not take notes because I was not thinking about notes. But I remember the tone, the edge in his voice. So the data that I have on these moments are what particular people wanted to tell me and not what I sought out or invited. I have heard only a dozen or so granular accounts in person.

I struggle with how to write about all of this from the infrastructure-rich environment of Wisconsin. The natural impulse, both for journalists and for scholars who value thick description, is to present first-person accounts, to pay attention to what this moment was like for particular people who were there when it happened. I am not so sure that this is what I can or should do. First, I do not want to beguile the reader with gripping, sensational accounts of disaster survival. Second, given that I was not there and only journeyed to Uttarakhand a year later, I do not want to write in such a way that the text persuasively collapses the insurmountable distance between what I write and what happened. I am distant from what happened. I am insulated. That is part of my position. Third, from the necessary distance at which I am writing, it seems to me that narration risks imposing chronology and structure on these minutes that maybe was not there. This assertion maps to one of my broader claims about the place of Kedarnath at other times, that it is a place where stories and words run out. So instead I am trying to write in a spare way that both suggests how things felt and does not pretend to know or even to offer the clearest possible picture. At the same time, I know that as I write I am lending scholarly weight to public perceptions of what happened at this time.

There are several digital animations on YouTube that include a computer-generated animation of the two flood events in Kedarnath. I have watched such animations several times, and though the shape of the computer-generated waves has a regularity that does not match the terrain it is nonetheless chilling. I had, and have, difficulty watching such animations. If you were in Kedarnath, then for several unrecoverable minutes this wave-event was all that was happening. There was a shattering sound. Then the mud-rock-wave began to fall down the slope. There was a moment to realize what was happening, but there was so much sound that it was without sound. You had to get out of the way. You had to help other people get out of the way. Everything narrowed. You either moved or did not. It was like this: _____.

Almost every account of this event that I heard a year later when I visited included some version of the Hindi verb *bhagna*, to flee. There were several ways to survive. All of them involved making a quick decision about where to run. The first was to seek high ground. If you were on the eastern side of Kedarnath, this meant heading away from the flow of mud and water toward the helipad or up toward the Bhukund Bhairavnath shrine and beyond. If you were caught on the path to Kedarnath, this meant scrambling up toward Vasuki Tal or above Garud Chatti and hoping the steep mountainside did not collapse. The path was collapsing and falling into the river. It was difficult to survive getting caught in the mud. When I visited the Kedarnath valley in 2014, an acquaintance of mine, with almost no change of expression, told me that he had been in Kedarnath at that time and had managed to save only one of his children from the mud. Reports of this sort were common, as were accounts in which the successful rescuer became the victim. Some strangers and loved ones saved each other in that moment of _____ and others could not.

Another impulse was to take shelter in the temple. The numbers conflict, but at least dozens of people went back into the temple. Shielded by a massive boulder, later termed a "divine rock" (Hindi: *divya shila*), and by its own solid construction, the temple itself held firm but filled up with mud and water. Some people also found themselves on the roofs or upper stories of buildings that did not collapse. If you made it up to the plateau behind the Bhukund Bhairavnath shrine, you would have gained enough distance to feel, temporarily, out of the situation. If you were in the temple, or on a roof, then you were marooned inside what was happening. And if you were in the temple, it is difficult to talk about later. All along the path down Gaurikund and beyond, people caught near rising mud and water sought high ground, and most (except in Rambara and Gaurikund) were successful. However, surviving the immediate event did not mean surviving the disaster.

Many people, once immediately out of harm's way, found a place to wait, and from that point it was a struggle against hunger and cold, often for several days, until relief came. This waiting took several forms. In the dark coldness of the temple it meant clinging for hours to something near the ceiling of the temple, like a bell, to keep one's head out of the muddy water and be one of those who would survive until the waters went down. This meant clinging to life for hours (and in some cases days) amid the bodies of the recently dead and dying (Press Trust of India 2013i; Rediff 2014).[3] Hridayesh Joshi (2014; trans., 2016), one of the first journalists who reached Kedarnath after the floods, offers a vivid, passionate, important account of this moment.

Visitors who had fled up the sides of the Mandakini River valley to the east or to the west then found themselves fighting hunger, cold, and trauma in a Himalayan environment for which they probably were not prepared under normal circumstances upon arrival in Kedarnath. The weather in the Uttarakhand Himalaya, even in summer, is different enough that in my experience most visitors (unless they are from a neighboring mountain-state like Himachal Pradesh or a country such as Nepal) are often not prepared because they do not think it necessary to purchase gear that they will need for less than twenty-four hours. When in Kedarnath I often felt, to my embarrassment, that without my costly warm and waterproof clothing and sleeping bag I would have found it impossible to live there. The same layers I wear in a Wisconsin winter were barely enough to be comfortable at night in Kedarnath in May 2014. Living and breathing at 3,500 meters also demands a massive caloric load. People who made it out of the floods had none of these things and simply had to wait for helicopters with soldiers to come because helicopters were the only possible means of delivering relief.

Others, mostly but not only locals, became convinced that it was not practical to await rescue and that it made more sense to walk to safety using high mountain paths that led into adjoining valley systems: if trapped on the western side of the Mandakini they went up and to the southwest beyond Vasuki Tal and then down to the villages of Tosi and then Triyugi Narayan, and if trapped on the eastern side of the Mandakini they went up past Bhukund Bhairavnath and then across and down to the southeast and eventually into the northern ends of the Kalimath valley system. This also took several days, and the villages deep in the Kalimath valley and above Triyugi Narayan saw groups of cold, starving survivors stumble in days later and immediately start looking for ways to let their families know they were alive. These survivors had made their way through cold jungles scattered with the bodies of those who had been too hungry, dehydrated, sick, or old to successfully walk to safety (Basu 2013, 90; HT Correspondents 2013; Zee Media Bureau and Press Trust of India 2013).

Those who did not find their way to a functional village with food and warmth also had to wait for several days before they were either picked up or in many cases given the short-term supplies they needed to stay alive. The weather was so bad that helicopters had difficulty flying up into the northern part of the Kedarnath valley. On June 21 a helicopter crashed during a rescue attempt. I do not know if it was the same helicopter, but I saw the remains of a helicopter between Gaurikund and Jungle Chatti on the path to Kedarnath in May 2014. Those whom helicopters did successfully pick up were first taken to staging stations farther south in the Kedarnath valley like Fata or Guptkashi or other sites near Rudraprayag like Agast-muni and Gauchar that had been high enough or far enough from a river to have come through the flooding with minimal damage and that had helipads. Eventu-ally the survivors moved from these places to Dehra Dun, where they, along with the other tens of thousands of evacuees, stayed until circumstances and resources permitted them to make their way home (Upadhyay 2013b). By this time survi-vor accounts were part of the public sphere across Uttarakhand, across India, and across the world (H. Kumar 2013a; Press Trust of India 2013d). Here is another summary, again from Piyoosh Rautela (2013, 47):

> The sequence of events in the Mandakini valley took everyone by surprise and none really got chance of raising alarm of any sort. Attempts were however made to communicate the news of flooding over high frequency police radio set but the seriousness of the incidence could not be assessed from hurriedly communicated incomplete message. All communication with Kedarnath valley was snapped in the late evening of 16 June 2013. Adverse weather and terrain conditions did not provide opportunity of resorting to alternative probes. The outside world as also the local administration therefore remained unaware of the events in Mandakini valley till the afternoon of 17 June, 2013.
>
> Aerial rescue operations could therefore be initiated in the morning of 18 June 2013. Ground search and rescue operations were hampered by washing off of the pedestrian track at many places between Gaurikund and Rambara and could start on 19 June, 2013. Terrain conditions made it difficult even to airdrop food and water at many locations where people were stranded in large numbers. Challenges faced even in aerial rescue operation can well be understood from the fact that three choppers got crashed during rescue operations in Mandakini valley.
>
> Massive ground and aerial search and rescue operations were however organized jointly by National Disaster Response Force (NDRF), Indo-Tibetan Border Police

(ITBP), Indian Air Force (IAF), Indian Army and Civil Administration to evacuate the survivors. Temporary helipads were quickly prepared and activated for evacuating persons stranded between Gaurikund and Rambara. In all around 6,817 persons were evacuated by air and another 18,183 by foot track. Despite best efforts evacuation could only be completed on 23 June, 2013.

For almost twenty-four hours no one knew exactly what had happened in Kedarnath, though the flood surge had obviously originated there. Then, despite full knowledge of what had happened, rescue efforts and helicopter reconnaissance could not begin until the next morning because of the combination of altitude, weather, and darkness. For almost a full day, there was simply no way to reach Kedarnath. The instability of the terrain and the continued bad weather made it difficult for up to two days even to deliver supplies to survivors between Gaurikund and Rambara, and attempts resulted in helicopter crashes. During those first days bodies were coming down the river and groups of survivors were coming out of the jungles and finding their way to villages in the upper Kedarnath valley and giving preliminary eyewitness accounts to media, to government agencies, and to local families. It was known that survivors of the floods were trapped and needed help, but the knowledge could not immediately be acted upon. But what this *meant* was that people struggling to survive on the ground occasionally saw helicopters but did not necessarily receive help from the helicopters they were able to see. The difficult location, combined with a profound lag in response time by the state government, meant that it took several days for the full impact of state and central government resources to be felt. In those early days, as Hridayesh Joshi (2016) has poignantly described, the gap was filled in part by commercial helicopter pilots and volunteers. For several days there was something very close to a silence, a gap. It took several days for those stranded in Kedarnath to leave the site by one way or another. It also took at least several days for anyone outside a disaster-struck location to obtain reliable news about what was going on. And then, like something you see first but cannot hear until moments later, trauma and confusion began to ripple out from those first moments. And the environment was not finished setting the terms. More bad weather was expected on the twenty-third, which meant that survivors needed to be evacuated before then (Gupta and Sunderarajan 2013).

THE BEGINNINGS OF THE AFTERMATH

By June 19, different threads of the *situation,* to use Jackson's term, began to ripple outward and separate out into different stories. There were rescue efforts, and the plight of survivors continued to unfold on the ground in the Kedarnath valley and in Uttarakhand more broadly. Rescues were effected by massive combined military civil rescue operations code-named Surya Hope and Rahat (India Today Agencies 2013; *Daily Bhaskar* Staff 2013). These involved the efforts of local civilians, police, the Indian Army, the Indo-Tibetan Border Police, the Border Roads Organisation, the National Disaster Response Force, the Indian Air Force—according to one news report the combined efforts of over eight thousand rescuers and fifty-five helicopters tasked with the rescue and evacuation of well over fifty thousand people (Gupta and Sunderarajan 2013). In these initial stages these prestigious, powerful resources of the central government took primary place, and even these resources were coming up against the limits of the situation—rescue helicopters trying to get to Kedarnath crashed during this time (Peer 2013).

At the same time, survivors were making their way out of the hills, as I mentioned earlier. Relatives of visitors to Uttarakhand who were missing had arrived in Dehra Dun and were showing pictures of missing family members to rescue workers coming down from the mountains. Growing numbers of visitors to Uttarakhand who had been rescued from immediate peril were arriving in Dehra Dun, sharing their stories, and on occasion being shown on large video screens so that they could be identified (Upadhyay 2013b; Chandramohan 2013b; BBC News India Staff 2013a; H. Kumar 2013a). By June 22, Kavita Upadhyay (2013b) reported that eighty-two thousand visitors had been safely evacuated from Uttarakhand but that over twenty thousand were still trapped in the Rudraprayag, Chamoli, and Uttarkashi districts. Also by June 22 the Railway Ministry had begun to run special trains for returning survivors to their homes.

Uttarakhandis themselves were realizing their losses, and in many places such as the Kedarnath valley were doing so amid electricity outages, roads blocked by landslides, interruptions in the arrival of goods from the plains, and a fog of uncertain and incomplete information about what was happening. Speaking on the phone with a friend who was in Ukhimath at this time was difficult: I imagined that I could feel something of the weight of the lack of electricity and diminished availability of fresh food combined with building grief and anxiety about the unfolding situation. At the same time my physical distance and safety from the situation made me feel almost dizzy. My friend said that people were spending a lot of time sitting inside and in the dark, full of grief and uncertainty, eating only staple foods like *dal* and *roti*. In May 2014 in Dehra Dun I spoke with an Uttara-khandi journalist who had journeyed via helicopter with the Indian Army and was one of the first civilians on the ground near Kedarnath. He spent four days walking between Chaumasi (north of Kalimath and southeast of Kedarnath), the remains of Rambara, and Tosi (above Triyugi Narayan to the southwest of Kedarnath), namely the area that at the time was filled with living, dead, and dying survivors of the floods. He gave food, assistance, and companionship where he could. "So much pain [Hindi: *dukh*]," he said to me, sitting in an overpriced coffee in the Barista coffee shop in Dehra Dun. "We are lucky [Hindi: *Ham bhagyashali hain*]. We should live life [Hindi: *Jina parta hai*]."

Announcements about financial relief began to circulate as relief supplies (food, clothing, kerosene, medicine) began to flow into the mountains. News articles and still and video images began to circulate on television, radio, the Internet, and Facebook and among cell phone users, celebrating the efforts of Indian solders to deliver food to survivors who had not yet been rescued and to help people cross swollen rivers to safe evacuation points (A. Bhatt and Press Trust of India 2013; H. Kumar 2013b; A. Bhatt 2013).

The materiality of death began to force its way still further into the situation. By June 21, bodies were washing ashore in Haridwar, over two hundred kilometers south of Kedarnath, and Kedarnath survivors were recounting the numbers of dead they had passed on their way out of the valley (BBC News India Staff 2013a; Press Trust of India 2013d). Google had created a Person Finder for Uttarakhand. As the number of confirmed dead rose, it was realized that many of the deceased could not be moved from where they had been found, and there began to be public discussions and disagreements surrounding the issues involved with a mass cremation in Kedarnath (Press Trust of India 2013b). The cremation, a public health necessity, was eventually carried out in Kedarnath after several challenges were surmounted. This was an immensely painful issue that added to the shock of loss because relatives did not have access to the bodies of their deceased family members within the required amount of time and could not honor their ritual obligations to the dead. There were public disagreements about the appropriateness and timing of the cremations and how they would affect the purity of the temple. When I spoke about this with a friend of mine in Ukhimath by phone on July 3, 2013, he said "He [Shiva] lives in the cremation ground, so what is the necessity?" (Hindi: Shamshan wala hai to kya zarurat?)

After being delayed by weather, the cremations began in Kedarnath on June 28. There were demonstrations in Dehra Dun where protesters demanded that the cremations not proceed (Bhaskar News Staff 2013). Protesters accused the state government of trying to cover up the true number of dead. I read about this online and heard about it in an e-mail from an incensed Garhwali friend at about the same time. He wanted me to write an exposé of what was happening. This was one of the first moments when I had to consider how my responsibilities were changing. I had become accustomed to resisting the push to publicly hymn the greatness of Kedarnath and say that I was writing a modern *mahatmya*, a hymn of glorification for the *tirtha*. At many moments during my fieldwork I spoke with someone who welcomed my presence because, it was proclaimed, I would return to America and undertake the promotion (Hindi: *prachar*) of Kedarnath. In most of these moments I would correct my interlocutor and say that my job as a researcher was not to do *prachar* but to write about reality (Hindi: *hakikat*)—what was really happening. This kind of statement made people nervous because in Kedarnath, as in any important place of religious and economic significance, there were gritty realities to how life was lived that did not fill locals with pride. I would explain that I did not intend my work to be an exposé that would focus only on the problems of the place but that I was attempting to give a full and truthful account.

But in 2013 there was a different tone to such questions—a demand from friends that I take a public position of critique, with the subtext "Here is something you can do that might actually help the situation, unlike most of the things you do in Garhwal, which are not relevant for us here." I said that I had begun to write something but that my sort of writing took a bit more time and that I was trying— an unsatisfactory answer. Even then I also knew that what I was writing would probably not be fully satisfactory to my friend. Edward Simpson, both in the context of both his own ethnographic writing on the aftermath of the 2001 Gujarat earthquake and in relationship to larger questions about disaster and memory, has noted that disasters and their aftermaths sharpen the issues of position inherent in ethnographic writing. Public opposition to his own work raised for him significant questions about how ethnographers should respond or change their work process if those about whom they have written seriously disagree with their conclusions (Simpson 2016, 2014; Simpson and De Alwis 2008).

An additional ritual uncertainty of primarily local significance was playing out at this time: How and where would Shiva in his Kedarnath form be worshipped with the temple closed? I was often told during my fieldwork in 2006–8 that the designated *pujari* of Kedarnath takes a strict vow to remain continuously in Kedarnath for the duration of the season and to perform the worship of the rock *linga* of Shiva in the temple and the *bhog murti* (a small portable *linga* that usually resides in the residence of the *pujari*) twice daily without fail. The *pujari* of Kedarnath and his assistant from the Samiti are in effect the only two people who simply cannot leave in a normal season. The *pujari* in 2013 was Vageeshvar Ling. He had come to Uttarakhand as an adult, and until his posting to Ukhimath had been from Karnataka and living in Karnataka. He and I were on the same bus when I journeyed to Ukhimath to begin my fieldwork. Vageeshvar Ling, in the midst of everything, found and kept the *bhog murti* safe. On June 18 he performed its worship in Garud Chatti, two kilometers to the south of the temple. Within several days he was in Ukhimath, where traditionally the *bhog murti* is worshipped as Shiva in Kedarnath form when Kedarnath is closed for the season. In 2013 the winter worship of Kedarnath-Shiva began at the end of June. This move was criticized by a North India–based guru, Swami Swaroopanand, and others who said that the worship in Kedarnath should have been continuous (Zee Media Bureau 2013; Khandekar and Nagrath 2013).

This criticism had a particular resonance in the Kedarnath valley because, as in all of Garhwal and many other Himalayan regions, deity journeys and the conditions of their journey are immensely significant (Halperin 2012; Sax 1991; P. Sutherland 1998; Berti 2009). Once when I traveled with the *doli* to Kedarnath, I was told that several decades earlier the Samiti had tried to take the *doli* by bus part of the way and that the result had been terrible storms. If part of the reason this disaster happened was that locals had not been properly respecting the resident divine powers, what would the effects be of breaking yet another tradition? Viewed from another angle, of course, there was no choice. Kedarnath was uninhabitable. Another event could happen at any time. It could not serve as a place of worship until this impurity had been at least, in some way, nullified. The entire upper valley needed to be fumigated from the air (*Hill Post* Staff 2013). After my visit to Kedarnath in May of 2014, workers were still finding new corpses on the path to Kedarnath (*Outlook India* Staff 2014).

Uttarakhand quickly became, at the national and state levels, a marked arena for political contestation. On June 20 the then chief minister of Gujarat, Narendra Modi, wrote the central government railway minister asking that extra trains be made available free of charge to return Gujarati evacuees home (Dave 2013). On June 22, when his request to visit the Kedarnath valley was denied by the Congress Party–led state government, Narendra Modi visited Dehra Dun and "did an aerial survey" of some disaster-struck areas to protest the reply of Home Minister Sushilkumar Shinde that politicians not make visits to disaster-affected areas because doing so would use valuable resources (*Hindustan Times* Staff 2013; Press Trust of India 2013c). Yet at the same time Rahul Gandhi, one of the central figures of the Congress Party, was permitted to travel to Gauchar, a village in north-central Garhwal with an airfield that was serving as a staging ground for rescue and relief operations (Paljor 2013). By June 24 there were already numerous public critiques of how the state government was handling the situation, many of which focused on the person of Uttarakhand's chief minister, Vijay Bahuguna (Aron 2013; BBC News India Staff 2013b). On June 30 Bahuguna rejected Modi's offer that Gujarat would rebuild the Kedarnath temple (Press Trust of India 2013a). Congress Party–led Uttarakhand emphasized its connection to the central (then also Congress Party–led) government. Yet at the same time Modi found an efficient and symbolically powerful way to deliver disaster relief up into the mountains very early on, at a time when officially involved governmental organizations were having difficulties doing the same. Gujarat sent small, well-packaged packets containing a series of essential items for short-term survival (J. Bhatt 2013). A friend (who like many Hindus in Uttarakhand was a strong Modi supporter) said that in Ukhimath the arrival of these packets was an impressive bright spot amid the logistical chaos of broader-scale relief efforts. At the level of state politics, the aftermath of the floods was an opportunity for persistent critique of the in-power Congress Party by the Bharatiya Janata Party (BJP) with the clear agenda of overthrowing Congress in the next round of state legislative assembly elections in 2017. By June 25 it was announced that rescue operations in Kedarnath were finished (Press Trust of India 2013j). One of the first news videos from Kedarnath, in which ABP News interviewed a representative of the Badrinath-Kedarnath Temples Committee, showed that just a week later Shiva's Kedarnath form had already begun to receive worship again in the simple form of *bilva* leaves, all other ritual materials being unavailable (ABP News 2013).

One of my main sources of information as events were unfolding was Facebook, particularly the Facebook networks of Garhwalis connected in some way or other to the Kedarnath valley. I saw cell phone and news videos (literally thousands of which were available on YouTube) of unfolding rescue scenes: soldiers carrying people up and down hills, helping people climb out of vehicles that had fallen off the road, feeding people, in some instances creating bridges with their own bodies. I saw many pictures of destroyed and submerged houses and roads. Many of these images were shot by someone from the area for viewing by other people from the area. I saw people trying to find out through Facebook whether a relative or friend had survived. One thread I viewed was written on June 19 by someone relaying (and translating from Garhwali to Hindi) and coordinating an appeal for help. The writer had a dying but partially functional cell phone and was trapped above Sonprayag with a group of about twenty people, several of whom were in bad shape and were not able to contact rescue personnel themselves. Digital photos of news articles were shared and reshared, many without searchable titles. Again and again in Facebook postings in these early weeks I also observed the anger of Garhwalis at the bad reputation (Hindi: *badnam*) being created by exploitative treatment of *yatris* and embarrassment that some of what was reported in the news might actually be occurring; in counterpoint, multiple postings stressed locals' generous and heroic behavior (*Hindu* Staff 2013b). In those first weeks *yatris* were getting most of the attention, and locals noticed. Here is an example of morbid humor (translated from the Hindi) that was posted around this time on the Facebook page of a friend of mine. It contrasts what the situation on the ground looks like to a reporter from a news channel compared to how it looks to a Garhwali:

> News Channel: Look how the soldiers are carrying people up these rocky paths on their backs.
>
> Garhwali: Brother, whenever anyone gets sick we carry them to the hospital like that.
>
> News Channel: Look at how *yatris* are saving their lives by traveling on these twisting paths. . . . And they are saying we will never again come to such a place again.
>
> Garhwali: Our children go to school on a thousand paths like this.
>
> News Channel: The rain still has not stopped—going any further will be dangerous. Our team has stopped here.
>
> Garhwali: Brother, if we waited for the rain to stop then we would never eat or drink, our children would not go to school, . . . our women would not go to work, our families would die hungry.
>
> News Channel: The road is broken here. . . . When the Indo-Tibetan Border Force comes they will repair it.

Garhwali:	Listen, younger brother, go get a shovel and a spade. The water is going to the wrong place. We will make a canal for the water and this will fix the road.
News Channel:	Look at how our cameramen and reporter have come by a difficult road and through cascading waters.
Garhwali:	Brother, on a day like this we have to cross cascading waters like this ten times a day. . . . Our barn is over in this direction. . . .
News Channel:	We have had with us for four days a local laborer—let's talk to him. "Are you connected to anyone who is trapped here and who assigned you here? Is the government paying you?"
Employee:	Sir, my name is Shishupal, and I am a commando in the Indian Army, enlisted in the Second Parachute Regiment. Just now I was on a vacation, and my home is here in Ukhimath. This tragedy has happened here so I know everyone involved; my uncle is trapped in Kedarnath, and I have come here to help him and find out his condition. This is my brother Shankar. He is a software engineer in Delhi; and is here now also on a vacation, and since the tragedy he has extended his vacation for seven days.
News Channel:	For some reason we will not be able to tell you the condition of the local worker [Shishupal] but we can tell you that the Bahuguna administration is doing good work and that local workers are working alongside the army.[4]

For locals, monsoon season in the mountains is not, for many, something from which there can be a rescue. Unless one moves down from the mountains and leaves Garhwal behind, mountain living in the best of circumstances is irreducibly hard. Further, in many cases local residents know their environment the best and are able to make the most useful interventions in the landscape. This point, discussed further in the next chapter, constitutes a refrain in literature about disaster prevention, disaster recovery, and sustainable development. And these facts of *pahari* life were constantly elided in the dramatic news pieces that showed people being heroically "rescued" by civilians, soldiers, relief workers, and other government personnel.

Imagine, then, how the early summer of 2013 felt to people in the Kedarnath valley. They were stuck in the middle of a heavy monsoon. Each day brought the threat of more landslides. People were in the first stages of dealing with great loss and in many cases were doing so without access to the bodies of their dead. They had ritual obligations that they could not fulfill properly. It was uncertain how much help was going to come from the government. Guptkashi (and to lesser extent Ukhimath, because of its location on the other side of the valley) was swarmed by survivors, relief supplies, nongovernmental organization (NGO) workers, journalists, and government officials. The state and central governments had begun financial assistance, compensation, and loan deferral for disaster-affected individuals and families but in an uneven way. There were numerous, conflicting announcements being made about the nature and time frame of reconstruction. There was no as yet universally agreed-upon explanation of the underlying causes of the disaster. And who knew if more was yet to come? It was still raining; there were still landslides.

If I force my mind away from the Kedarnath valley and think about the situation from the point of view of visitors to the region, it seems to me that the floods were, among other things, a complete betrayal of expectation about the nature of travel in the Himalaya at the beginning of the twenty-first century. When I spoke with people visiting Kedarnath in 2007 and 2008 about what they felt while present in Kedarnath, the answer would often include the words *peace* (Hindi: *shanti*) and/or *satisfaction* (Hindi: *santushti*). Some reported experiencing *shanti* because by virtue of being in a place like Kedarnath they had left all the worries of their daily lives behind (this was a very common sentiment), and *santushti* because they had successfully arrived in Kedarnath and by doing so had affirmed their connection to the older layers of Hindu traditions of which *yatra* was an important emblem. Added to this was the fact that in recent years travel in the Uttarakhand Himalaya, while still arduous and risky, had become increasingly comfortable—an attractive *yatra* destination for a family looking to get away from the heat of the plains, make a meaningful journey, and experience the famous "natural beauty" of the Himalaya, often while balancing the comforts of higher-end lodging with the mild arduousness of travel in the mountains. Thus the disaster was a profound betrayal, a shocking reversal of what a Himalayan *yatra* in the twenty-first century was supposed to be, the opposite of the feeling of rest, accomplishment, reconnection with tradition, and closeness to divinity/nature. For *paharis* it was a betrayal of a different kind, detailed in the next chapter.

REOPENING

Confusion swirled around the question of when the temple would reopen, who would be involved, and what would be revealed about the health of the physical structure itself. The Indian Army began work on a new footpath to Kedarnath from near Sonprayag. Building on the wisdom of most of the older travel routes in the valley, it would stick to ridgelines and avoid moving along the landslide-prone sides of the Mandakini River valley (Press Trust of India 2013h). Just months after officials had announced that it would be one if not several years before Kedarnath reopened, it was announced that Kedarnath would reopen on September 11, 2013 (Press Trust of India 2013e). Peer-reviewed work by scientists specializing in Himalayan geology and glaciology had also become a greater part of the public conversation about what had happened in Kedarnath. There had been numerous interviews and blogs on the topic beginning almost immediately after the floods (Bagla 2013; Mihir and Irani 2013; Petley 2013). But it took a bit more time for scientists, now working under a new kind of pressure and sense of responsibility for producing work that would enter public conversation, to offer preliminary analyses aimed at fellow scientists (Dobhal et al. 2013). The situation began to feel more of the weight of scientific modes of discourse and authority.

Kedarnath and the upper Kedarnath valley continued to feel like a place where human efforts were both going wrong and persistently attempted again and again, yet the pressure was on to rebuild Kedarnath so that the site could reopen as quickly as possible. The Rudraprayag-Guptkashi road closed several times because of landslides. Yet in the same month the Indian Ministry of Road Transport and Highways said that they planned to restore road connectivity to Kedarnath by the end of September (Press Trust of India 2013g). The Nehru Institute of Mountaineering recommended the development of an alternate route to Kedarnath from the east via Kalimath and Chaumasi (Gusain 2013b). Numerous NGOs were working with specific village populations (Dundoo 2013; Disaster Mitigation and Management Centre, 2015).

The BJP criticized the reopening of the temple on the grounds that it was a ploy to divert attention from state government mismanagement and that the association of Kedarnath pilgrimage priests had not been consulted (Chandramohan 2013a). Archaeological Survey of India and Geological Survey of India teams began mapping projects aimed at assessing the extent of the damage to the built environment and the broader geological situation. The opening precipitated a new round of survivor accounts and reviews of recent events in the media. The ritual dynamics of this time were fluid and confusing, a blend of the hopeful and the disturbing. Official worship of Shiva in Kedarnath began with a *havan* and *anusthan* performed by the Rawal Bheemashankar Ling that was attended by representatives of the Badri Kedar Samiti, the association of Kedarnath pilgrimage priests, and prominent state politicians. Significantly, weather prevented Vijay Bahuguna from attending as scheduled. Part of the work of this ritual was, in a ritual sense, a work of repurification. But this was a poignant repurification, as those present in Kedarnath continued to find new corpses as they removed layers of mud and debris from the ruins of the village. Locals were finding bodies as they excavated their shops and guesthouses. At some level, as Shashank Srivastava has suggested, these rituals of place and worship were aimed at self-repurification rather than repurification of place and temple.[5] On September 21 it was reported that the Archaeological Survey of India had announced that the *ghee* from the *ghee malish*, one of the distinctive features of a visit to Kedarnath in recent memory, had begun to rot and was preventing the stone from breathing naturally. According to *India Today*, this pronouncement produced a surprising response from the Samiti that avoided commenting directly on the *ghee malish* of the *linga* and instead focused on the walls of the inner sanctum: "Smearing the walls with ghee is a practice with no religious significance" (Gusain 2013a; Y. Kumar 2014).

On October 4, the first day of fall Navaratri, Kedarnath opened for *yatris*. This had traditionally been a time when Kedarnath saw many Bengali visitors. In 2013, the special connection between West Bengalis and the Himalaya was honored in a different way: Durga Puja celebrations in West Bengal saw the construction of many floats and displays depicting the Kedarnath temple. Actually reaching Kedarnath was difficult at this time. Worries about weather and landslides persisted, and even getting into the mountains felt like a gamble to visitors. Many places on the road between Rudraprayag and Sonprayag were dangerous. The government began requiring registration of *yatris* in Guptkashi and Sonprayag. The arrival of thirty-four people in Kedarnath on a given day during this period was newsworthy (*Amar Ujala* Staff 2013b). The state was trying to show that Kedarnath and the Char Dham were open again for business, but it was a difficult message to sell. On October 29 Kavita Upadhyay (2013c) wrote that "since June 24th 2013, 545 bodies have been disposed of and cremated in the Kedarnath Valley both, as part of the combing operations and debris clean-up. Almost 50 percent of these bodies were recovered from Kedarnath itself."

The doors of the Kedarnath temple closed for the winter on November 5. Raghav Langar, the district magistrate for Rudraprayag, announced that further recovery of the dead in Kedarnath would have to wait until spring (*Amar Ujala* Staff 2013a). Hundreds of people believed to be in Kedarnath still had not been located (V. Singh 2013). On November 6, Brijesh Bhatt (2013) wrote that 1,600 *yatris* had taken *darshan* since the *apda*. According to Ushinor Majumdar (2013), the number was "nearly 1,400." To put this in perspective, around ten thousand visitors *per day* usually came through Kedarnath during the high season in some of the years leading up to 2013. The Kedarnath *doli* returned again to Ukhimath after the traditional closing *puja* that is known as *samadhi puja*. As with the September opening, the procession took a slightly different route, skirting Rambara and Gaurikund and sticking to the higher footpaths, the thoroughfares of earlier times.

In Badrinath, as in Kedarnath, there is a tradition that when the shrine is closed it is the turn of divine beings to come to the site on pilgrimage while it is off limits for humans. With Kedarnath-Shiva in *samadhi* for the winter, it began to sink in that there would be no returning to how things were. Official statistics were updated and reconfirmed. On December 7, Kavita Upadhyay (2013a) wrote that the Ukhimath administrative block (Hindi: *tehsil*) had experienced 584 deaths and that 296 women had been widowed. High-level Uttarakhand government employees tasked with disaster recovery visited Orissa to compare notes. In October of 2013 Orissa faced inland flooding caused by Cyclone Phailin. In notable comparison to Uttarakhand, there were relatively few deaths because the state forced people to leave their homes (Press Trust of India 2013k). In what had now become a familiar postdisaster political trope, Modi continued to criticize the

Congress Party government in Uttarakhand. Worry over what would happen during the pilgrimage season of 2014 hung over the state. And it was a very frustrating worry because not much could be done on site or even beyond Gaurikund until the snows began to melt in the spring. But already Uttarakhandis were seeing that almost no *yatris* were booking rooms for the following season.

MOVING FORWARD

It has been observed that "disaster victims . . . tend to merge into an immediate unity that later fragments" (Hoffman and Oliver-Smith 1999, 7). This initial unity and later fragmentation were clearly evident in Kedarnath and in the Kedarnath valley, and the specific patterns warrant extended analysis. That is not, however, the focus I wish to emphasize here. Rather, I want to notice how the intersubjective unity of the immediate postdisaster moment nested inside the broader ways in which Kedarnath, as a Shaiva pilgrimage place in the Himalaya, generally creates and reflects shared experiences of the power of the place. That is to say, the disaster did not fundamentally question the commonly held idea and experience that Kedarnath is a place where a divine power is resident in and part of the place. Rather, it laid bare a destructive tangle of human relationships (with other humans, with divine powers, with the Himalayan landscape) that had, over time, become part of the complex eco-social system of Kedarnath.

In the first year after the floods, the Uttarakhand government found itself working at several different time scales all at once—trying to deal with immediate questions of relief, short-term questions about how the next year would unfold and how people would survive the coming months, and longer-term questions about what would be the new realities of the state. Journalists had written critically about the level of relief and rehabilitation the state had managed to provide by December (Majumdar 2013; Upadhyay 2013d; Press Trust of India 2013f). The tourism department had announced an array of initiatives aimed at both surviving the present and building the future: a new focus on adventure tourism that attempted to shift the tourist gaze away from the Char Dham, the regulation of the daily number of *yatris* in future seasons, the forgiveness or deferral of certain kinds of loans, and incentives for job retraining in the tourism sector (S. Sharma 2013; Majumdar 2013).

In the beginning of 2014, tensions between local groups, state, and center intensified, and confusion over how Kedarnath would actually be rebuilt deepened. The Geological Survey of India, an organ of the national government, stated in its initial report that nothing should be rebuilt in the immediate vicinity of the Kedarnath temple because of the instability of the area. They suggested (and pointedly referred to earlier reports where the same suggestion had been made) that the township be rebuilt slightly to the south and that the Mandakini River be rechanneled to its original preflood course. This immediately created an outcry in the Kedarnath valley community because local businessmen and particularly Kedarnath *tirth purohits* wanted their shops, guesthouses, and *dharamshalas* rebuilt (State Unit Uttarakhand 2016; Thapliyal et al. 2013; Gusain 2014b; S. Sharma 2014). Rahul Gandhi, a national public symbol of the Gandhian legacy in the Congress Party, met with the families of victims, and Uttarakhand chief minister Vijay Bahuguna urged Prime Minister Manmohan Singh to quickly supply national funds that could be used in road reconstruction (Press Trust of India 2014a; *Jagran Post* Staff 2014b). BJP criticism continued apace (Press Trust of India 2014c). Two days later, on January 31, 2014, Bahuguna resigned and was replaced by Harish Rawat, another senior figure in the Uttarakhand Congress Party who had narrowly missed becoming chief minister earlier (Press Trust of India 2014d). At almost the same time Uma Bharti, a highly visible member of the BJP at the national level, was placed in charge of the BJP unit in Uttarakhand (*Jagran Post* Staff 2014c). In the beginning of February the state received a massive loan of 1,200 crore (12 billion rupees) from the Asian Development Bank (*Jagran Post* Staff 2014a).

In one sense there were many encouraging signs—a change in leadership, massive infusions of funding, increased indications that the rebuilding of Uttarakhand would be a key issue between the BJP and the Congress Party at the national level. Harish Rawat immediately began working to reassure the public both inside and outside Uttarakhand that the state was on task (Upadhyay 2014b). He pointed out that much could be done under existing legislation without seeking new laws; designated the area between Rambara and Kedarnath as a "very special zone for rehabilitation and reconstruction"; said that there would be special safety apps for *yatris* and mobile health units stationed along the route; said that BSNL (the government communications company) would strengthen cell phone reception in the area; and promised that the route to Kedarnath and the Char Dham would be protected by round-the-clock weather satellite observation (Upadhyay 2014b; *Jagran Post* Staff 2014d, 2014e; Upadhyay 2014a; S. Joshi 2014b). On March 1 the state passed a resolution to compensate shopkeepers for inventory lost in 2013, extending the benefits beyond the simple loss of life and property (*Jagran Post* Staff 2014f). On February 27, the day on which the festival of Mahashivratri fell in 2014, *purohits* in Ukhimath calculated that the Kedarnath temple would open on May 4.

There was equal cause, however, for worry. Almost as soon as plans were announced for the 2014 season, anxiety began to build about whether Kedarnath would be ready to open (Press Trust of India 2014b; *Jagran Post* Staff 2014d). In the months immediately after the flooding, many had said the site would take years to reopen. The problem was that the state economy did not have years. Uttarakhand could not afford to not be Dev Bhumi, home of the Char Dham, for an entire year. The tourist industry was already on high alert by February because advance bookings for the upcoming season were almost nonexistent (J. Bhatt 2014). Not much could be done on the path between Gaurikund and Kedarnath until the snow began to melt (K. Singh 2014). In March, as would have been the case in any March, concerns were still being publicly voiced about the condition of the roads (S. Joshi 2014a). In the beginning of April new skeletons were found in Rambara, and Harish Rawat was not able to visit Kedarnath to assess progress because the weather was too dangerous for helicopter travel (V. Rautela 2014; Gusain 2014a; Kunwar 2014). Public voices were discouraging Char Dham Yatra booking (Vembu 2014). Reading these reports in Wisconsin, I did not know what to pack for my trip and decided to pack as if I would be camping in the snow in Kedarnath, which frightened me and made the trip feel like a foolish thing to do. I downplayed my fears to my family.

Uttarakhand was preparing for the opening of the season, and at the same time India was preparing for national elections. The reconstruction of Kedarnath and the rebuilding of Uttarakhand were an ongoing thread in the voices and thoughts of people all over the country. People who worked in the Kedarnath valley did not see any reason to make the investments they would normally make at the beginning of the season but for the most part were not finding workable alternatives. Rather, they were oscillating between anxiety and concern about what the state and central governments were going to do, grief for their losses, and a structural, collective depression that felt unsolvable. The state government needed to move as quickly as possible to rebuild and restore the hope and brand reputation of the state, but the complexity and severity of the terrain, weather, and logistics made this difficult. The new chief minister, Harish Rawat, was attempting to distinguish himself from his predecessor Vijay Bahuguna and to signal to Uttarakhandi publics that his administration would work in greater accordance with the ideals that had led to the establishment of the state (for example, he met with representatives from disaster-struck regions in Gairsain, the Chamoli district mountain town that many had thought should be the mountain capital of the new mountain-state of Uttarakhand/Uttaranchal). The state Congress Party knew that if the public was not satisfied with how Uttarakhand would rebound after 2013, and perhaps even if it was, after the state assembly elections in 2017 the state government might again find itself controlled by the BJP. The BJP would in fact go on to win a resounding victory in the 2017 state assembly elections. And beyond the drama of electoral politics, the government found itself needing to repair its relationship with its people and constructively reaffirm the *pahari* character of much of the new state. Hovering behind and underneath everything was the way that the landscape and the weather continued to push back and weaken human efforts. And the closer one came to Kedarnath, the more difficult and confusing everything became. This was the fraught situation into which I journeyed at the end of April 2014 when I traveled to Uttarakhand for the first time since the *apda*.

6

Nature's *Tandava* Dance

As a society we are not 100 percent blameless. Now the rivers are saying that
we are more powerful than your government, your planning commissions,
your real estate agents, and your contractors.
—SHEKHAR PATHAK (2013)

After I visited Kedarnath at the beginning of the season in 2014, it became clear to me that to understand reactions to the disaster (Hindi: *apda*) both inside and outside Uttarakhand it was necessary to understand the longer story of how, in discourses about and experiences of the complex eco-social system of Kedarnath, the following entities came to be experienced as enmeshed with each other: nature (Hindi: *prakriti*), Shiva-Shakti, and development (Hindi: *vikas*).[1] This chapter offers a set of vignettes that illustrate how this emergent enmeshing looked and felt in and around Kedarnath at the beginning of the season in 2014 and examines, building on the earlier chapters in the story of this place, how the persistently patterned enmeshing of Shiva-Shakti (particularly in Kedarnath) and local divine powers with different understandings of *prakriti* and *vikas* colored and created the conditions for these recent configurations. The complex interrelationships of humans, the divine, and the natural were deeply embedded in outrage and guilt about how recent policies relating to development and commercial infrastructure regulation as well as personal business decisions exacerbated the impact of the disaster in Uttarakhand and its immediate aftermath.[2] Pursuant to the broader aims of this book, my primary goal in this chapter is to situate Kedarnath at the beginning of the season in 2014 within the longer eco-social story of the place and the region and to give the reader a sense of experiential weight of all these forces as they combined and interconnected in Kedarnath.

Part of this longer story is the story of an at first gradual and then rapidly intensifying set of conditions through which Garhwalis, and those visiting Dev Bhumi, experienced themselves as suddenly vulnerable, or perhaps more precisely *vulnerable in a different way* than they had been in recent memory because their practices of living, building, and working were changing in a way that was out of step with the landscape. This approach to the *apda* in Kedarnath is consonant with the idea, characteristic of disaster studies since the 1980s, that what are often termed natural disasters (or sometimes "natural hazards") should be understood as events that reveal underlying, typically anthropogenic dissonances between particular populations and their environments that have arisen over time rather than sudden, surprising acts of ambush by the natural world (Gaillard and Texier 2010; Hewitt 1983; Oliver-Smith 1999a; Oliver-Smith 1999b). Further, the effect of a disaster on humans depends on the social, cultural, political, economic, environmental, and material contexts in which the disaster occurs. Specific people or groups, because of these factors, become more "vulnerable" to environmental events and experience the same event differently. As Michelle Gamburd (2012, 37) puts it, "Exploitative social structures that predate the event create conditions where certain groups suffer more during a natural hazard and have less capacity afterwards to recover."

In many Himalayan contexts, the incidence of floods and landslides has been increasing during the second half of the twentieth century, in part because of ill-planned construction of roads (such as the extension of the Kedarnath valley road beyond Guptkashi) connected to the Sino-Indian War in 1962 (Ives 2004, 93–95). In *Himalayan Perceptions,* a work summarizing decades of scholarship on Himalayan geology and environmental science first published in 2004, Jack Ives (2004, 118) wrote that "human interventions on the mountain environment frequently exacerbate pre-existing slope instabilities thereby accentuating vulnerability to mountain hazards."[3] *Mountain hazard* was a term coined in place of the more common *natural hazard* to emphasize specifically the way that "catastrophic processes" in the Himalaya, particularly glacial lake outburst floods (GLOFs) and landslides, were "augmented by human interventions on the landscape," especially the sorts of interventions created by "the continued expansion of modern infrastructure into a mountain system that only a few decades ago was isolated from world events" (Ives 2004, 133). These broad trends were explicitly connected to the increasing levels of "trekking tourism," whose impacts, in the Uttarakhand context, we can imagine would be similar to those of *yatra* tourism (Ives 2004, 140–41). Furthermore, GLOF events were known to "occasionally occur 'out-of-season'; for instance, after the assumed end of a monsoon season or prior to its onset" (Ives 2004, 135). This summary closely describes the Chorabari Tal flood event. In other words, anyone involved with mountain tourism in Garhwal could have looked at the Kedarnath situation and seen the terrible possibility that Kedarnath would eventually fit into these larger patterns. The floods of 2013 laid bare the failure of recent development trajectories and state governance because the floods and landslides themselves were not out of character for the region. It was the level of destruction caused in 2013 that was uncommon, along with the intensity of the rainfall and flow of water (H. Joshi 2016, 126–32). But much of the impact of this event was clearly and unsurprisingly human caused.

The fact that the state and central governments could have been far better prepared for the floods themselves and their multilayered aftermath was experienced as a deep betrayal of the public interest. The felt experience of this betrayal was vastly different for visitors and for Uttarakhandis. For visitors, the majority of whom were Indian, this stood as the latest in a series of recent disasters in India such as Cyclone Phailin (1999), the earthquake of Gujarat (2001), and the Indian Ocean Tsunami of 2004 that prompted the passage of the Disaster Management Act of 2005 and the creation of the National Disaster Management Authority. Flooding in Kashmir, Orissa, and Andhra Pradesh joined this list in 2014, and in 2015 the area in and around Kathmandu in neighboring Nepal experienced a devastating earthquake. The creation of this authority had been overdue—South Asia has seen dozens of disastrous events produced by floods, earthquakes, storms, and tsunamis just since the mid-twentieth century (Feener and Daly 2016, 14–21). Furthermore, in 2005, in part because of the massive impact of the 2004 Indian Ocean Tsunami, many Asian countries had become signatories to the Hyogo Framework, a "10-year plan to make the world safer from natural hazards" (UN Office for Disaster Risk Reduction 2007; Feener and Daly 2016, 25). Given these conditions, a modern twenty-first-century nation-state should have been able to provide advance warning of events like this and should have been prepared to hit the ground running in the aftermath, particularly in a region that had come to base significant parts of its identity and economy on different forms of mountain tourism.[4] Further, it should have included the inevitability of such events in its long-term planning. This did not happen in Uttarakhand in 2013. In the early twenty-first century the Uttarakhand Char Dham Yatra was beginning to feel modern and safe—a Himalayan family vacation that connected people with their traditional religious values and obligations. The region succumbed to one of the implied, and sometimes false, promises of modern built environments—that they can promise unnatural safety and predictability.[5] These floods, as journalist Shankar Aiyar (2012) has observed about disastrous events more generally in India, became a moment when the flaws of the nation-state became visible to all. Aiyar argues that in India these moments often have served as a catalyst for change.[6]

For Uttarakhandis the sense of betrayal and anger went deeper. The character and degree of recent infrastructure development in Uttarakhand exacerbated the death and destruction caused by these flood and landslide events. Coupled with the way that the plight of *yatris* immediately after the flooding received much more attention than the situation of locals, the situation felt like a betrayal of the ideals of sustainable, alternative development that had led to the establishment of the state of Uttarakhand. Shortcomings of the state's relief efforts further deepened this waking resentment. Thus, both inside and outside Uttarakhand, the progress of the state in finding a way through the monsoon and out of the disaster quickly came to function as an extraordinarily freighted, unavoidably public set of conversations about proper forms of development taking place at the collision of the ideal and the possible. On June 20, 2013, I stumbled upon a Facebook posting that captures this sentiment:

> Friends, you should know how it feels to be close to death. What it was like to have everything taken away on the sixteenth of this month, you should know this. On the morning after that night, the additional difficulties that happened to me, you should know this. On that dark night I was just praying to God [Hindi: *Bhagwan*] to never show that kind of night again. On the seventeenth it felt as if perhaps I would never get *darshan* of the sun again. In the entire Kedarnath valley ten or fifteen thousand people are missing, and *harv* rupees of goods have been scattered to the winds. People are trapped in one place. This is all happening because of dams— the glaciers became spoiled [Hindi: *kaccha*] from the digging of tunnels in the mountains. That is what happened in Kedarnath—a glacier broke Chorabari Tal, and from this far too many lives and much prosperity died. [The following switches to English] We have to protest the construction of Dams in Uttar[a]khand otherwise we have to be ready for this type of calamities again and again in Future.[7]

What I found notable about this wrenching comment is the decisive equation of the flooding with the building of dams and tunnels, written by someone who (presumably) less than three days before had been standing in Kedarnath. There is no uncertainty here. There is quick, furious certainty about a tragedy whose causes were, to this writer, already apparent within days of the flooding. On June 20, 2013, Shekhar Pathak, an Uttarakhandi historian and well-known public intellectual and environmental activist, offered an expanded and informed, yet no less passionate, critique:

> Twelve years earlier in Uttarakhand the building of roads, digging, sand extraction, electricity projects, etc. happened so fast, and in such an abnormal way, that the rivers decided to assume their dreadful manifestation. . . . They cast their vote about the current development prototype in which rivers are interfered with by dams, and tunnels are dug through mountains for the sake of electricity. Society repeatedly regarded the great government with hope, but in recent years the in-charge governments did not look at their own natural environment in a perceptive way. And society's members also interfered with the rivers everywhere, building hotels everywhere so that their own economic situation would improve. As a society we are not 100 percent blameless. Now the rivers are saying that we are more powerful than your government, your planning commissions, your real estate agents, and your contractors.

On June 25, 2013, Ravi Chopra wrote an editorial for the *Hindu*, entitled "The Untold Story from Uttarakhand," protesting the absence of any sort of public concern for the situation of Uttarakhandis themselves.

> While the focus is on pilgrims, nobody is talking about the fate of boys and men who came from their villages in the Mandakini valley to earn during the yatri season. . . . Last week's disaster not only spelt doom for thousands of household economies but also dealt a grievous blow to Uttarakhand's lucrative religious tourism industry. With the media focus almost exclusively on the fate of pilgrims, the scenes of the deluge and its aftermath will linger on in public memory, making the revival of tourism doubtful in the foreseeable future. The abject failure of the State government, political leaders and the administration is therefore likely to impoverish the State coffers too.

Journalist Hridayesh Joshi, one of the first observers to reach Kedarnath after the floods, wrote an entire book about the Kedarnath floods that has subsequently been translated into English (H. Joshi 2014, 2016). His carefully researched reports offer a trenchant summary of the governance failures:

> Two months after the disaster, a senior IPS officer involved in relief work in the Kedar valley told me: "We were not at all prepared for an emergency like this. In such a critical situation, we worked very slowly. . . . We were also thoroughly inefficient. No one had any idea what to do. Trouble had started on 15 June itself. . . . Major destruction followed on 16 and 17 June . . . but till 19 June we were sitting around, twiddling our thumbs." . . . There was an inordinate delay in pressing in the army and the air force into action, resulting in a tremendous loss of lives. However, once the army and the air force began rescuing people, there was a visible improvement in the situation and it is largely due to this reason that many lives were saved. (H. Joshi 2016, 98)

There is a sense of outraged critique present in these pieces that both expresses and describes a fundamentally regional sense of betrayal. Uttarakhand was intended to be a new state that, in a departure from the recent history of the region, would value the long-term needs of its residents and the health of the environment over and above what would benefit or make sense to those outside the state. Many of the problems that exacerbated the unavoidable destruction caused by this event reawoke long-standing resentment about the ways in which first colonial and later postcolonial political and economic structures systematically exploited the natural resources and labor of mountain (Hindi: *pahari*) people. But there was a new tone here as well. The mismanagement of *yatra* tourism and connected infrastructure projects was a new addition to this set of regional grievances that, given the state of preexisting knowledge about sustainable mountain tourism, could easily have been avoided. This is a key point for understanding what led up to the floods of 2013—that the new practices and infrastructure supporting *yatra* tourism were in important ways continuations of older patterns in the region formed through the intersections of colonialism, globalization, and specific visions of development. Resentment against *yatra* tourism was not culturally marked in the same way as the building of dams and the cutting of trees. After 2013, however, the older regional *pahari* sentiment fueled this new conflagration as the trauma of betrayal pulled everything into a tighter focus. The vulnerabilities produced in the aftermath of the floods of 2013, vulnerabilities that yet again disproportionately affected the *paharis,* for whom there was no airlift or rescue from the situation, were in a sense a return to earlier forms of *pahari* resentment that had animated protests against logging and hydroelectric dams. This time, however, the resentment had a more inchoate target—a set of socioeconomic conditions that had encouraged the Garhwali *paharis* connected to *yatra* tourism to invest, build, and act in ways that they themselves knew were a departure from traditional Garhwali lifeways.[8] The "natural" resources being exploited were not something physical like forests or water but rather something less tangible—the partially commodified power and beauty of the Uttarakhand Himalaya and, in the case of Kedarnath, the added drawing power that came from the place-specific presence of Shiva and Shakti and other resident divine powers.

VISIT TO KEDARNATH

Almost a year after the floods I flew to Delhi. Over several days, using a combination of travel by train, bus, and jeep, I made my way into the Kedarnath valley and caught up with the opening procession in Guptkashi. May 4, 2014, found me struggling up the path from Gaurikund to Kedarnath, the last leg of the procession. The *doli* carrying the traveling form of Kedarnath-Shiva had already passed me and I could not keep up. I spent the last chunk of the ascent, especially from Linchauli onward, in the occasional company of two state government employees whose jobs required that they work in Kedarnath. They were coming to Kedarnath for the first time. One said that he felt a bit scared when he found out he would have to live in Kedarnath for a month. I later met a third state government employee who had survived the *apda,* had come back in October of 2013 with the relief and reconstruction teams, and now found himself here again. He said he was going to ask for a transfer the next day because it was too hard for him to be there. He had survived five days after the floods by staying near the Bhukund Bhairavnath shrine on the upper east side of the valley before being picked up by a helicopter. When he returned to work around a week later his appearance and affect were so different that his coworkers did not recognize him. It turned out that a renunciant (Hindi: *sadhu*) I had met often by the Kedarnath temple who also survived the disaster had fed him milk during those five days. My conversation partners kept waiting for me to catch up because I was moving so slowly, was clearly very tired, and was having difficulty catching my breath. Later they told me they had been a bit worried about me.

I stumbled into what the Garhwal Regional Development Authority (GMVN) publicly terms a prefabricated structure, what I would call a tent. I woke up the next morning too late for the first round of opening ceremonies, which was noted by one of the young priests employed by the Samiti with whom I had sat and walked in previous years. On the one hand, I felt embarrassed that I did not have the mental fortitude to force myself out of bed earlier. On the other hand, I reminded myself that even making it to Kedarnath had been a great challenge and I simply had to do what I could. I later realized that part of what was going on was probably pent-up shock and grief. At around 6:30 in the morning I went to the temple. The doors had opened earlier for VIPs, but at that time the doors were being formally opened for *darshan* and *puja* for the general public. Rawal Bheemashankar Ling, the guru for Kedarnath *pujaris,* addressed the media as a press of people waited to enter the temple. I could not hear what he said because I was not ready to take my boots off and thus could not enter the temple courtyard. I felt unequal to the moment.

Once the doors had officially opened I got ready to enter the temple myself. I made myself bathe—I borrowed a bucket, stripped down, and did a quick bucket bath in the open air behind a building to the north of the temple in the freezing air. I went inside with the flower that a shopkeeper on the Gaurikund-Kedarnath path had given me to offer in the temple along with a few flowers I had gathered. I felt ashamed because they had become crushed from being in my backpack. In comparison, I had seen a *sadhu* who walked the entire route holding a metal canister of perfectly placed flowers in his hand. The day before I had seen several men walking up the path holding a single unblemished flower in one hand the entire way. When I entered the inner sanctum and came near the *linga* there is only one way to express the moment: I was "overcome with feeling" (Hindi: *bhavuk ho gaya*). A year of pent-up grief came out the moment I touched the *linga,* as if I had been holding my breath for a year without my knowing. After a time a pilgrimage priest gruffly indicated that I needed to conclude my cathartic moment so that others could have their turn. Several Kedarnath locals saw this happening, and now I suppose it is a part of what is known about me.

I spent the day wandering around Kedarnath village—looking, listening, talking. I ate lunch courtesy of the Shree Kedarnath Sewa Mandal, a group that since the floods has made a multiyear commitment to feeding people in Kedarnath. A journalist from West Bengal asked me for a comment. She wanted to know my reaction to the fact that so many people had come to Kedarnath in spite of the tragedy of the previous year. I said I thought people were coming both in spite of the tragedy and also because of it, which nonplussed her. She wanted me to affirm her formulation that Kedarnath today had become a symbol of faith (Hindi: *astha ka prateek*). While I think this sentiment is in many ways descriptively accurate, I felt uncomfortable at being invited to affirm this formulation because I felt that it left out too much about what was happening. What we were seeing was also about economic, political, and social pressures.

I spoke with a *sadhu* I recognized from years past. He told me about how he had survived by sprinting in the direction of Bhukund Bhairavnath in the morning

FIGURE 3. The Kedarnath temple on opening day in 2014.

FIGURE 14. A side view of Kedarnath temple on opening day in 2014 with the garlanded *divya shila* just behind the temple and the remnants of the rock–debris-flood leading back up to Chorabari Tal in the background.

after hearing the cries of the dying in the temple all night. The *sadhu* said that Shiva has turned him into a real *sadhu* by destroying all of his possessions. He talked about the numerous corpses that must still be hidden all through the village underneath the rubble. He called Kedarnath a "city of ghosts." He told me that Mahant Chandragiri, the renunciant who had run a food kitchen at the northern edge of Kedarnath for many years, had died during the first flood event. I watched *yatris* and locals exploring the field of newly strewn boulders by the temple, both photographing and worshipping the gigantic boulder (now termed in Hindi *divya shila,* divine stone) that had miraculously prevented the floods from destroying the temple and was now becoming an object of worship and spectacle in its own right. I looked up the valley side to the north and saw, to my amazement, that some visitors had climbed up past the Bhairavnath shrine and were sledding down the hill. This struck others and myself as quite inappropriate, in several ways. First, it was an example of the kind of irreverent tourism that some people feel may have had something to do with Shiva's anger, and the hilarity felt out of place. Second, it was not yet appropriate to go near Bhukund Bhairavnath until the official Bhairavnath-*puja* that opened the season, which had not yet happened.

Later that day I had a conversation with an employee of the Uttarakhand Department of Tourism. I asked him whether he thought the government should be thinking about a different kind of tourism. He agreed but noted that the state was so dependent on tourism and it was such a fickle industry that Uttarakhand residents could not practically make the kinds of changes that they might prefer because they needed to make every effort to make things easy for visitors. He also said that although what had happened here had received a disproportionate amount of attention because of the spectacular circumstances and the number of outsiders involved, landslides and floods were a normal part of life in Uttarakhand. By way of example he noted that there had been bad landslides in Ukhimath in 2012. Part of what motivated his comments was that he was from the Uttarkashi area, which had had its own share of tragic floods and landslides, particularly in the aftermath of the construction of the Tehri Dam and most recently after the floods of 2013 (Drew 2014a).[9] He understood that events in Kedarnath had attracted public (and my) notice but wanted to be clear that for him the events themselves were not unique to Kedarnath or discontinuous with mountain life in Garhwal. He said that the policy makers in Dehra Dun were trying to shift Uttarakhand tourism from being solely Char Dham focused but that it would take time, money, and long-term political will. He mentioned an initiative to turn the lake created by the Tehri Dam into a tourist destination for water sports as an example of what they were hoping to do.

During that day in Kedarnath I found myself struggling between simply giving my own personal sense of shock, loss, fear, and awe full rein and forcing myself back to my responsibilities as a researcher to try and learn as much as possible while I was actually present in this difficult-to-reach place. In the early afternoon I was sitting in the temple courtyard, whose pristine condition (the result of months of labor that had involved the removal of meters of mud and dozens of dead bodies) was in marked contrast to the snow-covered and mostly ruined condition of the rest of the built environment. This contrast was a telling indication of the *shakti* of the place and, as the local man Shukla-Ji had expressed to me (in the Introduction), of the ways that this *shakti* could be understood as both inherent in the place and a result of human experiences and interactions. The temple and immediate environs looked this way because they had been the focus of postdisaster reconstruction efforts, more so than the rest of Kedarnath. But the temple itself had received this disproportionate amount of attention because it was the point where people felt the importance and power of the place to be centered. Later, I would describe Kedarnath in this way to people in the Kedarnath valley who asked me how the place looked: "Shiva is fine, everything else is in bad shape" (Hindi: Baba theek hai, baki buri halat). After saying this I would watch to see if my formulation made sense. Most of the time it seemed to scan and no one corrected or contradicted me. After trying this sentence out a dozen or so times I started letting the second half of the sentence dangle: "Shiva is great. Everything else? . . ." When I did this, several people filled it in in this way: "Shiva is fine. Everything else—useless/without meaning" (Hindi: Baba theek hai—baki bekar). While I was in Kedarnath, and afterward, several *Ved-pathis* and *tirth purohits* told me that *yatris* who had been to Kedarnath before would arrive in Kedarnath distraught at the level of ruin, would enter the temple, and would come out proclaiming that inside everything felt the same and that it had been an especially good *darshan*.

A conversation sprang up between myself and several Kedarnath valley locals. One man asked me if I thought Kedarnath could still function as a place on which one could base a livelihood. I became the focus of several intent gazes, each belonging to someone who had himself been in Kedarnath during the flooding, lost loved ones, and survived. I said I thought yes but not the way it had been before. One man said that it was the job of the government to return things to a better version of how they had been. For development (Hindi: *vikas*) to return, the government needed either to rebuild a motor road up to Gaurikund or to extend a road even further to Linchauli (halfway between Kedarnath and Gaurikund, the new midpoint that replaced Rambara on the eastern bank of the Mandakini) so that normal business could be restored. He proclaimed that the more *yatris* came to Kedarnath the more the place was spoiled (Hindi: *bigar jana*). This echoed the most common phrase used as an explanation for the floods that I heard from people in the Kedarnath valley: the floods were, in some way, a result of people "making Kedarnath into a picnic place," a place with in-room hot water and the option of different regional cuisines, environmental pollution created by thousands of people throwing away one-use plastic raincoats every day and other problems of sanitation, and relaxed expectations about appropriate behavior. However, the speaker continued by unapologetically (and to me, surprisingly) asserting that a certain degree of interference (Hindi: *cherchar*) with the place was necessary if local people were to survive. At the same time, however, several months after the floods the Geological Survey of India had recommended that nothing be rebuilt in the immediate environs of the Kedarnath temple because of the instability of the ground. Kedarnath pilgrimage priests and business people who were not part of the government but who possessed many powerful political patrons vehemently opposed this recommendation because it struck at the heart of their livelihood and their traditional rights, and they made it known that they would resist the government on this point.

The relationships of specific Kedarnath valley locals to the state and central governments were fraught and complicated at this time. It was the job of the state and central governments to help them, but help of that level necessarily involved regional and national politics, and it became clear that Kedarnath was being drawn further into larger-scale political narratives. For example, Kedarnath was in the process becoming a reconstruction project of national importance that was the special concern of the man who would become the new prime minister, Narendra Modi. Kedarnath has since been included on the list of sites that will be part of the national Pilgrimage Rejuvenation and Spiritual Augmentation Drive, or PRASAD (Press Information Bureau 2015). Rahul Gandhi, one of the national faces of the Congress Party, made a trek from near Gaurikund to Kedarnath by foot at the beginning of the 2015 season. On the other hand, Kedarnath pilgrimage priests and local business people were *pahari* Garhwalis who resented the assertion of any kind of control from outside the valley. Kedarnath had become, even relative to other disaster-affected sites in Uttarakhand, a distinctive point of intersection between what Arun Agrawal and K. Sivaramakrishnan (2003, 21) call the "high modernist projects of the Indian state" and a new, postdisaster iteration of what they term, as discussed in chapter 3, the "regional modernity" of Uttarakhand.[10] Many Kedarnath locals wanted to retain as much autonomy as possible. In this overdetermined arena, issues such as the construction and rebuilding of roads, the regulation of pilgrimage tourism, the construction of commercial and private buildings, and the control of space were intensely symbolic and intensely contested (Cederlöf and Sivaramakrishnan 2006, 9; Coleman and Eade 2004; Coleman 2002; Eade and Sallnow 1991). Brian Larkin (2013, 333) has observed that "infrastructural projects" often function as the placeholders that allow people to "participate in a common visual and conceptual paradigm of what it means to be modern"; "Roads and railways are not just technical objects . . . but also operate on the level of fantasy and desire."

All of this was hanging in the air as we sat in the newly cleaned temple court-yard, remembering what had been, wondering about what would be coming next.

Most of the people in Kedarnath on that day had been there the year before. The air was heavy with everything from memories of the past year to the presence of national media to the activity of the special task force composed of a number of different state, national, and private agencies involved in the work of reconstruction. The traumatic memories of what had happened jostled for space with political, economic, cultural, religious necessities that had forced the doors open again so soon and attracted people once again.

I felt that I had not done a good job of this conversation because I had been, at some level, trying to act like a researcher gathering data. So when I wandered over to one of the few buildings besides the temple that was still standing where several employees of the Samiti were staying I tried to inhabit the conversation in a different way, letting go of the idea that I would get information and simply trying to be present in a more human way. I did not try to conceal how emotional the visit was making me, and I think I saw the same thing on their faces. When we started talking, the idea came up that this was Shiva's place, not meant for people, something I had been told many times during my earlier fieldwork. One man, looking at me with the face of someone who had been in Kedarnath a year before and survived, said three words: "nature's *tandava* dance" (Hindi: *prakriti ka tandav*).

This phrase caught my notice. The more typical phrase would be Shiva or Rudra's *tandava* dance (Hindi: *Shiv ka tandav* or *Rudra ka tandav*). This is the dance Shiva dances to destroy impediments to liberation, to destroy in order to make space for a new cycle of creation, to rhythmically express his creative power. It is the dance Shiva danced to conquer the destructive being Muyalakan produced by the egoistic sages in the Pine Forest. It is the dance through which the divine totality shifts between emission and absorption, the play that makes space for the world. As Don Handelman and David Shulman (1997, 152) put it: "The world as seen from, or through, Śiva's dance is one which is being, or has already been, sucked back into the infinite density of the whole."

But the priest substituted the word *nature* for Shiva. I have since thought carefully about this moment and have seen and heard similar phrases repeatedly. It has become clear that this phrase offered an important window into thoughts and feelings about the disaster because it illustrates the core assertion of this book—that the idea of eco-sociality is key for understanding how people conceive and experience, pre- and post-*apda,* the overlapping relationships between the putatively separate categories of human, divine, and natural that animate the Shaiva *shakti* of Kedarnath. One need only survey headlines about Uttarakhand flooding to see examples: "Monsoon's *Rudratandav* [Tandav dance of Rudra/Shiva] in Uttarakhand" (*Hindu* Staff 2013); "When the Ganga Descends" (Padmanabhaan 2013); "When the Himalayas Poured" (H. Kumar 2013b); "Facing Nature's Wrath" (Bhandari and Benson 2013); "How Uttarakhand Dug Its Grave" (Mazoomdaar 2013); "Flood Fury: Writing Was on the Wall" (*Down to Earth* Staff 2013). The agentive subject shifts from monsoon to Shiva to Ganga to the humans of Uttarakhand to nature. The idea of *tandav* pulls the overlap of these elements in a Shaiva direction. The shifting agencies and linkages between human, divine, and natural on display here (which, it should be remembered, need not be understood as ontologically distinct categories) underscore the eco-sociality of the system.

I attempted to test my understanding of this conceptual complex on my journey back down to Gaurikund on May 5, 2014. This was difficult because I spent most of the walk focused on my own physical ordeal. It started to rain soon after I left Kedarnath, which meant that many parts of the path, which was still under construction, became coated with mud. Around Linchauli the soles of my boots separated from the rest of the boot, so I switched to sandals. The next several hours were mostly a step-by-step descent down a trail of mud and mud-covered stone, and my primary concern was getting down without injury. Despite the conditions I did manage to have an important conversation. I spoke with a group of five men from Delhi, one of whom came every year. I asked the one who came every year why he thought this had happened. First, he gave a Shiva-oriented explanation. He said that Shiva had become displeased with the state of the place. Then he said that there was, in addition to a *dharmic* explanation, a scientific explanation that these sorts of things happen every so often. The way he said that this made me wonder how my identity was affecting the way he was responding to the question. While I cannot be sure because it was a brief interaction in trying physical and weather conditions (a common problem with informal interviews conducted on the path between Kedarnath and Gaurikund), it felt to me that he did not want me to think that he was unaware of the scientific explanation, or that when he offered the Shiva-oriented explanation he felt pressure to also offer a corrective. I tried out the *prakriti ka tandav* formulation and he agreed with it. I asked the group how many people would come from Delhi this year. They said: not very many. The oscillation between and mixing of *dharmic* and scientific causation on display in this conversation reminded me of how Gamburd (2012, 35), in Sri Lankan conversations about the causes of the 2004 Indian Ocean Tsunami, found that "the power of karmic dynamics" and "geophysical explanations" were identified as (sometimes overlapping) causes, with informants preferring the "geophysical explanations."

Upon my return to Ukhimath several days later, I again explored the *prakriti ka tandav* theme. In a long conversation with several employees of the Samiti in their offices across from the Omkareshwar Temple, I invited input on how to formulate the relationships between Shiva and *prakriti*. After a good deal of back and forth I came up with the following, which was approved by those present: "We can understand Shiva's power by looking at nature [Hindi: *Prakriti ko dekh kar ham Shiv-ji ki shakti samajh sakte hain*]. This is because Shiva caused nature to tremble [Hindi: *prakriti hila di*]." The reason the temple in Kedarnath survived the floods was that Shiva can cause *prakriti* to tremble but *prakriti* cannot cause Shiva to tremble. We made a distinction between three elements: Shiva, *prakriti,* and *shakti.* Shiva and *prakriti* are different/separate (Hindi: *bhinn*). However, in another conversation with a Kedarnath *pujari* a bit later on the same day he offered a different formulation: "*Prakriti* is the true form of *shakti*" (Hindi: Prakriti shakti ka svarup hain). This is a different formulation that pulls Shiva into the flows of water, rock, mud, and debris that descended on Kedarnath. It also recalls, implicitly, one of the most famous stories about *shakti,* water, and Shiva in Hindu traditions: the descent of the Ganga. Shiva decided to stop filtering Ganga's power for a brief moment. Shiva is both connected to and separate from *prakriti,* and *prakriti* is a term that can signify both the natural world and under certain conditions particular forms of the Goddess, especially those connected to flowing water. This means that to understand these reactions and formulations we must not only understand the social and conceptual history of how people in different times and places in South Asia have understood the relationship of Shiva and Shakti, discussed in chapter 3, but also briefly investigate understandings of the term *prakriti* and its partial English cognate *nature.*

PRAKRITI

The idea of what twentieth-century Western thought might term nature (Hindi: *prakriti*) has a complicated and diverse history in South Asia both inside and outside the contexts of what some might today term explicitly or implicitly religious worldviews. In my discussion I alternate between writing *prakriti* and *nature* to draw the reader's attention to how many modern South Asian understandings of *prakriti*/nature are in a sense hybrid concepts whose ancestries are both South Asian and European. To acquire a broad sense of some of the possible nuances of the term *prakriti* in the phrase *prakriti ka tandav*, it is necessary to first briefly move back in time and then rejoin the history of Uttarakhand from the point of view of the Kedarnath valley presented in chapter 4 as the story moves into the decades just before the floods.

As David Haberman (2013, 44) has put it, the idea that the entire world constitutes an organic "manifest" "divinity" in which all beings are interrelated is a common religious idea in South Asia with a long history. *Shakti* and *prakriti,* along with terms such as *brahman* (Sanskrit: ultimate or underlying Being, Existence), *qudrat* (Hindi/Urdu: divine power, creation), *jiva* (Sanskrit: life, life-spark), and *dharma* (Sanskrit: cosmic architecture, fundamental particle, balance, justice, and many related meanings), are terms that, in the broadest sense, frame this understanding. Different South Asian systems of thought have understood the nature and permanence of this underlying relatedness in many different ways. Contemplation of the natural world can, under the proper conditions, lead to liberating knowledge (Chapple 1998, 30).

Prakriti is an old term. Along with the Sanskrit term *purusha* (primordial person, consciousness), it is one of the two entities from which all reality proceeds in Samkhya, a philosophical system with roots over two thousand years old. K. L. Sheshagiri Rao (2000, 25) puts it this way: "*Prakriti*, cosmic matter, is the matrix of the entire material creation." *Prakriti* can also be understood as a form of *shakti*. *Shakti* is a form of the Goddess who/which is often understood to contain and embody all goddesses and is often linked closely to Shiva. Today *shakti* is an idea that links the Mahapuranic worlds of Durga and Parvati, physical understandings of phenomena such as heat and fire, the local worlds of village goddesses (Hindi: *gram devi* or *gram devta* for village deity more generally) and understandings of the personhood, nature, power, and authority of women and, to a lesser degree, men (Padma 2013, 207, 267–68; Flueckiger 2013, 269; Shulman 1980; Wadley 1975). Both *prakriti* and *shakti* are also sometimes connected to the idea of *maya*. *Maya* can be both creative, the power through which the world becomes manifest, and "delusive . . . unreal," the illusion from which arises ignorance about how things truly are (Pintchman 1994, 93–94). *Shakti* also connects to both sides of what Tracy Pintchman (1994, 202) has termed the "intermediacy of the feminine," which oscillates and transitions between the auspicious and the inauspicious, the life-giving and the life-destroying. Forms of the Goddess, local and region-ally important goddesses, and place-specific forms of *shakti* constitute many of the primary threads in the fabric of Hindu sacred geographies. Anne Feldhaus (1995) has noted the deep connections that bind together women and rivers in the state of Maharashtra. The historical identification of *shakti* and the Goddess with the natural world has served as one of the most important resources and at the same time problems (because of the equation of the natural with the feminine) for Indian environmentalists and ecofeminists (Sherma 1998, 96–97). Vandana Shiva (1988, xvi–xviii, 38–40), who posits the harmonious cooperation of humans with the natural world as an antidote to regnant patriarchal models of development and agro-technology, uses this kind of understanding as the basis for her work (see also Chapple 1998, 23).

In South Asian contexts some relevant and important conceptions of nature did not emerge directly out of the conceptual categories that crystallized in epic, Puranic and tantric, and philosophical Sanskrit texts. Relationships with the natural world, as they inflect through agriculture, hunting, material forms such as textiles, performance, ritual, and metaphysics, have been of profound and ubiquitous importance for those tribal and indigenous peoples in South Asia whom Madhav Gadgil and Ramachandra Guha (1995, 3) term "ecosystem people," or those who historically "depend on the natural environments of their own locality to meet most of their material needs." The worldviews of such groups, the groups whose local goddesses and gods sometimes became incorporated into the Shaiva, Vaishnava, and Shakta frameworks of the epics and *Puranas*, were not fully (or often even partially) represented in Sanskrit-based genres and conceptual categories but have nonetheless been influential in how human-nature relationships have been understood and experienced in South Asia broadly and in the central Indian Himalaya specifically.

In the Tamil-speaking regions of South India, a famous set of classical poetic conventions established correspondences between particular interior emotional states and particular kinds of external natural landscapes that have deeply influenced how people in Tamil-speaking regions have lived their geography since the third century CE and have shaped the character of Tamil devotional poetry and sacred geography through to the present, which has in turn influenced other regional cultural geographies beyond Tamil-speaking regions (Ramanujan 1967; Selby and Peterson 2008, 1–16). Mughal and Persian understandings of the beauty of the natural world, alternatively, stressed the importance of agentive human involvement with the natural world. Christopher Hill (2008, 65–66) observes that the "practice of equating nature with heaven," particularly through the construction of gardens meant to invoke "Paradise on Earth," was a common and important way of picturing and relating to nature in medieval South Asia (see also Inden 2011). The domestication of nature through the construction of royal gardens became a political strategy for showing that the king held divine favor and became a theme that would influence the depiction of landscapes in "miniature painting," an important collection of artistic genres in medieval South Asia (Hill 2008, 66).

During the periods of colonial encounter and British rule, the demands of trade and empire (and particularly the building of the railroad system) discussed in chapter 4 overlaid onto these numerous, interconnected layers of geographic imaginaire an instrumental conception of the natural world as resource and commodity that functioned on a massive scale. Among the many effects of this change was widespread deforestation across the subcontinent. Other effects included a shift in agriculture toward the cultivation of monoculture "cash crops," higher levels of land taxation, and the production and extraction of "cheap raw materials"—"biological produce, rice and cotton, jute and indigo, tea and teak, as well as gold and precious stones" under deeply exploitative labor conditions (Gadgil and Guha 1995, 9–11). As already discussed in chapter 4, the instrumental and exploitative use of natural resources in South Asia by the British hit the Uttarakhand region hard and catalyzed popular movements that protested logging and the construction of hydroelectric dams (Gadgil and Guha 1993, 222–24; 1995, 22–24, 72, 84–85).

NATURAL BEAUTY

In the later stages of the colonial period it is possible to see constructions of "nature" by South Asian intellectuals that integrated and reinterpreted earlier streams of thought and practice with ways of imagining the incipient nation-state. One example was how poets belonging to the Chayavad movement of Hindi poetry deployed the idea of nature. Valerie Ritter has charted how the evocation of "Natural objects" blended the nascent Indian nationalism of the late nineteenth century with a "furtively" recovered poetics of the *shringara rasa,* the Sanskrit dramatic modality of eros and love, to create a "natural state of *śṛṅgāra* in the Indian media." Ritter (2011, 248) argues that this use of nature imagery, with its mix of new nationalism and old aesthetic conventions, laid the basis for what would become a profoundly important popular idea, often referenced in Indian popular culture, of "India's freedom itself as if a bride." The connection of territorially grounded forms of nationalism to the body of the Goddess is, of course, best known in the concept of "Mother India" (Hindi: *Bharat Mata*), which makes the territory of the nation-state a form of the Goddess.

When I would speak to visitors to Kedarnath who did not present themselves as first and foremost participants in a devotional *yatra,* they would often say that their primary motivation was to see the natural beauty (Hindi: *prakritik saundarya*) of the Himalaya. This contemporary usage in part derives from how the Himalaya began to be viewed as "picturesque" during the eighteenth and nineteenth centuries. The European idea of the picturesque mountain landscape began to circulate in India as European explorers, photographers, colonial administrators, and soldiers trekked, traveled, painted, photographed, and wrote about the high Himalaya. Indian painters learned to paint in "Company" styles influenced by Western notions of landscape, perspective, and portraiture (Tolia-Kelly 2007; Tillotson 1990; Kennedy 1996; Whitmore 2012). The British began to build hill stations. The increase in travel to *tirthas* and fairs seen during the second half of the nineteenth century extended to the Himalaya, and by the beginning of the twentieth century the semantic range of *yatra* started to include forms of trekking and leisure travel that in the West would come to be termed "tourism." Sandeep Banerjee (2014) has described the emergence of the idea of the Himalayan picturesque by showing how the underlying structures of British colonial power created the conditions for the domestication of the Himalaya through both the construction of highly controlled, racially segregated hill stations that evoked the English countryside and photographs of Himalayan vistas carefully framed to avoid the representation of local people and to evoke comparison to the European Alps.[11]

Banerjee and Basu (2015, 616) have demonstrated that this notion of Himalayan natural beauty carried multiple associations that changed over time for particular groups, notably Bengalis. They observe that early nineteenth-century Bengali travelogues imagined a "Himalaya infused with an inner spirituality" reflective of the "monotheistic synergies between Sufi Islam, and Vedantic Hinduism" found in the work of Rabindranath Tagore. The authors make the case that in the second half of the nineteenth century urban middle-class (Bengali: *bhadralok*) Bengalis traveling to the Himalaya shifted from conceptualizing it as "a sacred space" to seeing it as "a spatial metaphor of a putative nation-space," reflecting a process of "secularization" and the emergence of a civil religion of the nascent nation-state of India (Banerjee and Basu 2015, 609).This recalls the uses of nature by the Chayavad poets described by Ritter, but in a Himalaya and Bengali-specific way.

NATURE IN UTTARAKHAND

Many of these broader ideas about nature/*prakriti* came to inform how people in Uttarakhand today and those who visit Uttarakhand view the natural world and conceptualize their relationship to it in environmental, cultural, political, and religious terms. It is, however, also important to attend to regionally specific understandings of *prakriti* that come into play when we think about "*prakriti ka tandav*" and *shakti* in and around Kedarnath and in the central Indian Himalaya more broadly.

Over the last century and a half, and particularly in the last twenty to thirty years, ideas about and aspects of nature/*prakriti* have come to function as marked terms, "a major and explicit point of reference" in understandings of what after 2000 could be called "Uttarakhandi" regional identity or more broadly *pahari* identity (Linkenbach 2006, 152, 164). For Uttarakhandi *paharis*, according to Linkenbach (2006, 164), "The hills are geographically and ecologically set apart from the mainland, they are rich in natural resources such as forest and water, they provide pure air. As a *dharmic* land, the hills are responsible for the religiosity and honesty of the inhabitants; they provide a morally pure space."[12] To be *pahari* and from Uttarakhand was not to reject the encroachment of modernity and development but rather to envision a specific kind of development and "regional modernity" that were in line with the ecological and moral patterns of the mountains (Agrawal and Sivaramakrishnan 2003). This move was a cornerstone of the Gandhi-inspired Sarvodaya social justice activist movement and also of the philosophies of the Chipko movement introduced in chapter 4 (Moore 2003, 187). It represented an intersection of "regional modernity" with what Cederlöf and Sivaramakrishnan (2006) term "ecological nationalism."

Thus the idea of nature/*prakriti* operative in the Uttarakhand of recent decades connects the worlds of forest, field, river, and mountain to several different modalities of geographic experience. On the one hand the natural world is infused with local and translocal manifestations of *shakti* and the presence of locally, regionally, and nationally significant divine agents. Networks of locally and regionally significant deities, spirits, and ancestors pervade the natural landscape in a manner that is similar to that described by place-based indigenous (Gadgil and Guha's "ecosystem people") traditions in most parts of the subcontinent and more broadly across the world (Sax 2009, 54). On the other hand, the idea of nature evokes discourses of nationalism, regionalism, political economy and ecology, and development (Hindi: *vikas*). Seeing the involvement of the idea of *prakriti* in all of these related contexts helps us approach the feeling that, in the aftermath of the floods of June 2013, somehow all of these interrelated spheres of human activity and experience contained a sense of wrongness.

HONA HI THA

On the way back down from my trip to Kedarnath at the beginning of the season in May 2014, I spoke with a Kedarnath valley friend, as I had on the way up, in one of the few places that had not been destroyed in 2013. He had been standing on the roof of a building in Kedarnath when the floods came, and he survived. We spoke about everything from why I lived so far away from my parents to how people had been doing since the *apda* to the nature of my ongoing research about Kedarnath. We discussed what I was planning to write. Eventually he made a point of letting me know a detail that he considered very important for understanding how people in the Kedarnath valley viewed the *apda*. He said there was a phrase that almost everyone used when talking about it: "It was bound to happen" (Hindi: Hona hi tha). When he told me this, I noted that talking about the disaster in this way left ambiguous whether "it was bound to happen" because the Himalaya was a change-able natural environment that had to be approached carefully, because Kedarnath had "turned into a picnic place" and Shiva noticed, or because proper building codes, site administration, and advance warning had not been present. He agreed that it was ambiguous and left it at that. In a later return to this conversation he emphasized that he had been mostly referring to the physical likelihood of a flood event of the kind that happened in 2013, suggesting to me an uneasiness about what I might have taken from our previous conversation.

Several days later, in Ukhimath, another Kedarnath valley local said about the disaster, "Everyone has this *dosh*" (Hindi: Prani matra doshi hain). *Dosh* is a common term in Hindi and many other South Asian languages that derives from the Sanskrit word *dosha*. *Dosha* and its cognates in different South Asian languages, depending on context, can mean criminal guilt, "bodily disorder," (for example, in the Ayurvedic sense), an astrological problem, a family problem, "fault," or a curse or affliction originating from a god, demon, or spirit (Sax 2009, 37, 59, 114; Allocco 2009; McGregor 1993, s.v. "dosh"). People, soil, crops, and animals can have *dosh*, and there are an array of individual and collective practices through which *dosh* can be lifted. I was told in Kedarnath several times that when Garhwali deities came with people from their villages on a *yatra* that was not part of an established *yatra* cycle (such as, for some Garhwali *devtas*, going on pilgrimage every six or twelve years) there was probably some sort of collective or familial *dosh* behind the deity's visit and that the visit to Kedarnath was for the purposes of erasing the *dosh* (Hindi: *dosh nivaran*).

I read this post-*apda* statement that everyone has *dosh,* then, as a statement that can be analytically thought about at many different levels: as an internal disorder in the eco-social fabric of the valley; as the displeasure of a deity at specific actions indicating a lack of respect; as the product of the untimely death of many people; as the emergence of a disjunction between people and place; as the changing and souring of social relations that went hand in hand with the recent hypercapitalization of the valley.[13] It might also mean that lifting the *dosh* would somehow involve all or most of these different factors because, to return to the argument of the chapter, the divine, the natural, and the socioeconomic are bound up with each other in powerful ways in Kedarnath. This line of interpretation is a way of analytically conceptualizing the situation in ways that are in dialogue with ways of thinking on the ground in the Kedarnath valley. Radhika Govindrajan (2015, 504), in the context of fieldwork on forms of "interspecies kinship" in the central Indian Himalaya, describes a profoundly salient conversation with Devidutt, a Kumauni man, about what caused the Kedarnath floods that also connects to this idea of *dosh.* Devidutt rejected the idea that the floods were caused by anything having to do with ill-planned construction or weather changes produced by climate change. Rather, for him they were clearly the result of "divine revenge" against the 2011 Uttarakhand High Court ruling that had prohibited animal sacrifice, itself a substitution for the earlier practice of human sacrifice. In his view, "The gods had resumed taking human sacrifice and would continue to do so until animal sacrifice was offered to them again."

If one puts these two sentiments together ("Hona hi tha" and "Prani matra doshi hain"), then a very important understanding emerges about how some Kedarnath valley residents who were involved with the pilgrimage-tourism industry have felt since the *apda.* There is a sense both that they are victims of an event in which divine agency, geological and meteorological processes, and economic patterns were all involved and that they are also personally and collectively somehow responsible. What happened was an event of natural/divine causation. It was somehow the product of the rapidly expanding bubble of economic growth and commercial infrastructure development of the last decade and a half, the result of processes that were both transparent and opaque to the people involved. It was the product of people not acting as they should with each other, with the divine powers of the valley, and with the natural world. What happened was an intensification of what always happens in a Himalayan landscape prone to landslides and floods, but that intensification could have been prevented. Now we have to lift the *dosh* because it is our *dosh.*

VIKAS

Part of the *dosh* has to do with the idea and practice of "development." The wrong kind of development (Hindi: *vikas*) exacerbated the severity of this disaster (Jayaraman 2013). *Development* is a global term with many different context-sensitive meanings that typically have to do with the improvement or ostensible improvement of material, somatic, social, economic, cultural, and political living conditions. This idea has often functioned as the placeholder for what it was thought formerly colonized and other non–Global North peoples had to do to become successfully modern according to the standard set by the Global North. Haripriya Rangan (2004, 207) has offered a useful summary of what this term signifies in a general sense in India:

> There the idea of development (referred to in the vernacular as *vikas*, meaning both the process of moving towards the dawn of a new social era, and the social era itself) has been used as a secular, democratic means for opening the political arena to the claims of various groups in civil society. Development is charged with the promise of change towards greater social equality and prosperity for all citizens. It has been taken up by disadvantaged groups as a means of gaining political recognition and access to economic empowerment.

This general understanding has acquired specific meanings in Uttarakhand in recent decades. Pampa Mukherjee (2012, 201) has shown that an "alternative vision of development" was formative in the political visions for the new state that emphasized sensitivity to local needs and local autonomy in "ecologically sound" ways. For example, Mukherjee has observed that there was a strong push to shape Uttarakhand into a preeminent producer of organic produce and "floriculture" because this was a kind of production that could be carried out in localized, decentralized terrace farming that would be possible and sustainable in the mountains (212). However, to a large extent this was not the kind of growth that happened after the creation of the state in 2000. Instead, industrial manufacturing fueled strong economic development in those sections of the state not located in the hills (Mamgain 2008; Kar 2007).

In the mountainous terrain constituting the majority of the state, a good deal of economic growth was greatly influenced by the needs of different kinds of tourism, some more related to *yatra* and some less. However, the growth of infrastructure and the nature of economic development connected to pilgrimage and tourism and more broadly in the mountainous regions of the state did not unfold in a sustainable direction. Rather, to return to the voices of outrage with which the chapter began, the spatialization of capital took a path contrary to the founding vision of the state. Most of the building and investing in the Garhwal of recent decades was aimed at servicing the needs of the increasing numbers of middle-class *yatris* and the North Indian power grid without regulatory regard for the long-term environmental health of the region and its human and nonhuman residents. Hridayesh Joshi (2016, 133) again provides a useful and trenchant summary: "Despite a series of disasters in Uttarakhand, the government has no specific policy for development and planned construction keeping the environmental issues in mind. . . . Since the state leaders themselves are involved in hospitality and real estate, both overtly and covertly, no one actively discourages illegal construction." Very little of what happened with water and earth in June of 2013 was unprecedented. The terrible surprise was the personal, social, structural, and political myopia that made the impact of the floods and landslides so severe. It is because of this longer framework of causes and conditions that it makes sense to characterize the floods of 2013 as, in important ways, an *unnatural* disaster.

VIKAS BEFORE THE *APDA*

When the term *development* (Hindi: *vikas*) came up in conversation in connection to Kedarnath during my predisaster visits to the Kedarnath valley, it usually had several specific and linked meanings. An old woman from Jaipur, in a conversation in Kedarnath on May 24, 2007, linked *vikas* to amenities (Hindi: *suvidhaen*)—the availability of rooms with good blankets, warm water, and different kinds of food, telephone, electricity, and medical care. For several pilgrimage priests with whom I had conversations on the topic in 2007 and 2008, *vikas* meant increased access to Kedarnath via helicopter, the motor road, and other potential transport avenues such as ropeways. Implicitly, these were the avenues through which most people whose livelihood was connected to Kedarnath were earning more, borrowing more, sending their children out of the Kedarnath valley (in some cases as far as Dehra Dun) to go to school, enjoying the fruits of disposable income, and becoming reasonably confident that their children would enjoy greater access to health care, education, and eventually salaried employment (Hindi: *nowkri*).

Vikas in this sense was something that, prior to 2013, almost everyone in the Kedarnath valley even remotely connected to the pilgrimage tourism industry wanted to encourage. More and more middle-class *yatris* were coming who were willing to pay for higher-end food, rooms, textiles, and conveniences. It made sense to borrow money from banks and relatives to service this growing need.[14] It made sense for roads to be widened without regard for their long-term stability. In 2011, friends of mine in Kedarnath were adding new rooms and floors to their lodges and *dharamshalas* that targeted this demographic—rooms with in-room hot water heaters, a rooftop "cafe" with a nice view. In 2007 there was a "coffee shop" on the main street in Kedarnath selling instant coffee for fifteen to twenty rupees a cup. While the Samiti did build two large structures where large groups of poor *yatris* could, at a pinch, spend the night, relatively little concerted attention was being given in pre-2013 Kedarnath to the building of large, inexpensive accommodations.

For many locals in Kedarnath, in Gaurikund, and throughout the Kedarnath valley, discussions of *vikas* immediately became discussions about avenues of transport to Kedarnath: motor road, helicopter, horse, and ropeway. The necessity and/or imminent extension of the motor road to Kedarnath or Rambara was a common topic of chai-shop discussion in Kedarnath. Many locals were strongly in favor of this because it directly equated to an increase in the number of visitors.

During such discussions, however, it was often pointed out that people have been lobbying to bring the motor road to Kedarnath for decades. I remember mentioning this once to a Garhwali professor, and he immediately and vehemently said that there should never be a motor road to Kedarnath even if it meant that Garhwal would remain forever poor. When I repeat this story I am always quick to point out that his income was not tied to Kedarnath.

The tension generated by conversations about the motor road paled in comparison to the conflict that could be generated by the subject of helicopter *yatra*. Over the course of the previous decade, helicopter *yatra* had become an increasingly common feature of daily life in Kedarnath, particularly in the high season. When you were walking or riding up the path from Gaurikund, at many spots the helicopters zoomed up the valley pass very close overhead because of the relative height of the path and the narrowness of the valley. I remember thinking that it was a very strange juxtaposition between a traditional mode of traveling to Kedarnath for over a millennium and a form of transport that was very new. Many *yatris* would stop walking to point or take pictures. Sometimes it was the first time people had seen a helicopter close up. Kedarnath valley locals in the upper half of the Kedarnath valley whose businesses were not connected to helicopter *yatra* often resented the growth of helicopter service because it meant that their rooms, food, and shops were being overflown by the middle-class *yatris* with more money to spend. Others thought that this kind of resentment was shortsighted because eventually the Kedarnath valley pilgrimage economy would reorganize itself and the horse porters and lodge owners would be able to make money doing something different. At the beginning of the 2007 season I witnessed a demonstration against helicopter *yatra* in Gaurikund, and it has been a matter of considerable tension in Kedarnath. There was a strike by horse porters in Kedarnath about helicopter *yatra* in Kedarnath immediately before the floods in 2013.

While this is my analytic imposition, I have come to feel that the god Bhairavnath is understood to be, at some level, anti-*vikas*. Certainly he is anti-built environment. Bhukund Bhairavnath, the resident deity for Kedarnath whose job it is to protect the area, has been repeatedly asked to permit the building of a proper temple over his shrine in Kedarnath, and he has refused, saying, "I need to be able to see in all directions" (Hindi: Mujhe charon aur dekhne ke liye chahiye). I have also heard that this is the view of the deity Jakh who lives above Narayan Koti in the Kedarnath valley and whose Vaisakhi *mela* is one of the biggest events of the year.[15]

THE TONE OF RECONSTRUCTION

All of these tensions could be seen in how reconstruction in Kedarnath had been proceeding when I visited in 2014. I noted the profound ambivalence about how to proceed that often marks disaster-affected regions in the aftermath (Feener 2016).

The founding vision of Uttarakhand needed to be visibly upheld in the reconstruction process. The needs of residents had to be publicly valued. The state was theoretically aiming for a slower, safer, and more responsible kind of growth—carefully planned, sustainably developed *yatra* tourism that could regulate, limit, and reframe the flow of visitors to the Char Dham sites and Hemkund Sahib and instead invite them to visit other places of natural beauty and cultural importance in Uttarakhand. For example, in a meeting with an official in the Department of Tourism for Uttarakhand in May 2014, he said that Uttarakhand tourism recognizes that the number of visitors to Kedarnath (and hence the Char Dham) is likely to decrease in the near future (as it has, though it significantly rebounded in 2016) and that they are working on schemes that will encourage other forms of tourism—adventure tourism such as watersports on Tehri Lake or tourism where you stay in someone's home and take walking tours or view folk performances. His goals were to make travel to Uttarakhand feel "safe" and to decenter, to the extent possible, Char Dham Yatra tourism. It was very clear in 2014 that this kind of limited, diversified, and decentralized tourism could be beneficial for the region in the long term. Such plans were attempting to achieve a more "holistic" kind of economic development while not jettisoning the idea that Uttarakhand needed a great deal of revenue from tourism to flourish.[16]

These plans were not a wholly acceptable solution for people whose lives were involved with Kedarnath because any sort of thoughtful regulation of the area was likely to reduce the number of visitors, diminish local autonomy, and reduce the possibility of extremely high earnings in any given short period of time. There was, and continues to be, a sense that the goal is to attract as many visitors as possible to Uttarakhand for *yatra* tourism and that it is the responsibility of the state and central governments to help. In the Kedarnath valley the drawing power of the *yatra* industry even made it difficult to convince locals of the importance of integrating sustainable natural resource management techniques into current agricultural practices (Maikhuri et al. 2014). On the other hand, Kedarnath pilgrimage priests and local business people often resented state control or regulation from outside the valley. Kedarnath pilgrimage priests had already been in conflict with the Samiti, which is partially connected to the state government, for years about issues relating to side-door entry into the Kedarnath temple and saw postflood reconstruction proposals as extensions of this conflict that furthered abrogated their traditional rights and their Kedarnath-based livelihoods. Did the resistance of the Kedarnath pilgrimage priests to the Geological Survey of India proposal mean that they were blind to the necessity for reconstruction in Kedarnath that would be geologically, architecturally, and environmentally responsible and that they were unaware of their own role in how Kedarnath had changed in recent decades? It did not. The situation was simply more complicated than that.

THE *TANDAVA* OF A COMPLEX ECO-SOCIAL SYSTEM

Undergirding this welter of social, political, and economic contestation was the broader ecological experience of and uncertainty about climate change, deforestation, earthquakes, the increased incidence of landslides, floods (particularly GLOFs), and the impact of anthropogenic landscape changes such as road construction and widening and the hydroelectric dams whose construction generated the outraged sentiment quoted at the beginning of this chapter. The floods of 2013 and the destruction caused by flood, landslide, and earthquake events in recent decades can all in one way or another be connected to each of these issues. These complex connections are sometimes termed "hazard cascades" (Gardner 2015, 354). Specifying the exact nature of the connections is difficult. Climate scientists in recent years have begun to spend more time looking for connections between specific weather events and longer-term patterns of climate change, but these connections remain fuzzy in the absence of sufficient longitudinal data and measurement techniques that have earned scholarly consensus.[17] Events similar to the floods that struck Uttarakhand in 2013 are becoming more frequent because of climate destabilization, and discourses about climate change now inform how Uttarakhandis themselves think about what they are experiencing (Mathur 2015a, 88). GLOFs will probably become more frequent. Nayanika Mathur (2015a, 87) has described how in recent years climate change talk was deployed in the Indian Himalaya both as an "explanation for recurring incidences of human-animal conflict and the disappearance of a protected species through the labors of the local state bureaucracy" and as an example of a "conspiracy" that aimed to draw a veil over the real problem. In one case, locals in the Gopeshwar area told her that in recent years bears had been going mad and becoming increasingly aggressive because they, like humans, were angered by "the barefaced exploitation of the natural resources by the state and corporations working in tandem" (99–100).

The degree of blame that could be attributed to the construction of large hydroelectric projects, one of the targets of accusations quoted earlier in this chapter, was in 2014 similarly fuzzy. In April of 2014 a group of researchers headed by Ravi Chopra submitted a report to the Ministry of Environment and Forests of the Government of India entitled "Assessment of Environmental Degradation and Impact of Hydroelectric Projects during the June 2013 Disaster in Uttarakhand" (Chopra et al. 2014). The commission (given on August 13, 2013) for the constitution of the expert body and their report arose out of a preflood court case involving a hydroelectric company: *Alaknanda Hydro Power Co. Ltd. versus Anuj Joshi & others* (Chopra et al. 2014, i). The purpose of the report was to investigate what impact existing hydroelectric projects had on the effects of the flooding in June 2013 and more broadly on environmental degradation in the region and to consider whether data gained from the 2013 floods might suggest a different regulatory policy for existing and proposed hydroelectric projects in Uttarakhand. The necessity of such a report was obvious given the sentiments expressed at the beginning of this chapter. The report, over two hundred pages, concluded that twenty-three out of twenty-four existing and proposed large-scale hydroelectric projects posed significant impacts to local and regional biodiversity and that these projects should not proceed or be allowed to function at full speed until full environmental impact reports had been submitted and reviewed. It further noted that there were competing narratives about the impact of the dams on the floods. Reports furnished by the hydroelectric projects argued that some dams (such as the Tehri Dam) had held back the full force of the floods.[18] Chopra and his team of his researchers (2014, 10) offered a nuanced counterargument that argued for a closer examination of how dams managed muck or sediment flow, particularly in the case of the Phata-Byung (in the Kedarnath valley), Singoli-Bhatwari (in the Kedarnath valley), Vishnuprayag, and Srinagar hydroelectric projects (see also H. Joshi 2016, 146–47). Much of the destruction caused by the floods in 2013 was caused not by pure water but by muck: water infused with rock, dirt, sand, and mud that had built up and then overflowed barriers and boundaries. However, this report did not correlate the floods themselves to the building of hydroelectric dams. Rather, the report upheld the urgency and the complexity of the question of the degree to which human activity had worsened the impact of the floods. To summarize, understandings, theories, and claims about the degree of human responsibility (and specifically the humans residing in the region) for the flood and its impact displayed wildly contradictory variations that implicated the state and its residents to varying degrees. It remains unclear in a general sense where exactly to locate the building of hydroelectric dams inside this broader landscape of anthropogenic landscape change in the Himalaya, except for the obvious hazards created by building, in the name of *vikas,* large hydroelectric projects in a seismically active region (Valdiya 2014).

Since 2000, people had come to experience the *vikas* that came with *yatra* tourism, and that was itself connected to this longer and broader history of *vikas* phenomena in the region, in several different ways. *Vikas* was problematic because it had begun to turn the *tirtha* into a picnic place and had diluted the purity of life in the mountains. And at the same time *vikas* was linked to the influx of money that was quickly raising (by some measures) the standard of living, the process enabling people to connect to national and global economies to a far greater degree than ever before. The vicissitudes of pilgrimage tourism in Kedarnath and the Kedarnath valley before and after the *apda* came to be linked to discourses about development and discourses about the relationship of humans with the various forms of divine power embedded in and constituted by the natural environment signified by the terms *Shiva, shakti,* and *prakriti.* The idea of *prakriti* functioned to bind together this cluster of discourses, experiences, and practices.

In many ways the interconnectedness of this Kedarnath-based system of Shiva, *shakti, prakriti,* and *vikas* recalls what Ann Grodzins Gold (1998, 178) and Bhoju Ram Gujar noticed about how Rajasthani villagers considered morality, landscape, weather, human relationships, divine anger, and deforestation to be all part of "One Story." Gold and Gujar found a perception of an underlying unity and connectedness of living things that framed discussions of environmental degradation. In the broadest sense, environmental degradation was understood to be in part produced by the ripening of human *karma* (174). In Kedarnath, post-*apda* reactions and opinions about how reconstruction should proceed reflected these same conceptual linkages. People proclaimed that Kedarnath need-ed to become less of a picnic place (out of respect for *prakriti,* for Dev Bhumi, for Uttarakhand, for the sacredness of a *tirtha*) and that the redevelopment of the place should proceed along sustainable lines. Uttarakhand needed to rebuild Kedarnath in a way that would be in accordance with the famous pronouncement of Sunderlal Bahuguna that "ecology is permanent economy" (James 2013, 1). All of these requirements pointed toward the necessity of an overall long-term decrease in the number of visitors to Kedarnath.

However, during my visit to the Kedarnath valley in 2014 it seemed that the idea that pilgrimage tourism should not be the main avenue for economic devel-opment there was almost unthinkable. The force of the market was pushing in the opposite direction, making Gold and Gujar's "One Story" a story that could be told only under certain conditions. Kedarnath valley locals did not feel just one way about why the *apda* happened and how reconstruction should proceed. In her book on Ganga-related environmental activism, Georgina Drew (2017, 54) reports a conversation with Dr. Ramaswamy Iyer in which he insightfully and simply expressed a similar complexity regarding how people think about the Ganga by stating that "a river is many things." Drew accordingly observed a complex and connected set of understandings regarding the causes, effects, and remedies for how climate change and the construction of hydroelectric dams such as the Tehri Dam affect the flow of the river goddess Ganga. She also found, in separate but related research, that people living along the banks of the Ganga in and around Uttarkashi both "acknowledged the possibility that glaciers such as Gangotri-Gaumukh may be melting because of our global emissions and environmental hubris" and "at the same time" felt that the Ganga could survive these changes because she also existed in the "heavens" and the "underworld" (Drew 2014b, 34 35). Additionally, "Hindu conceptions of right action" informed people's understand-ing of the anthropogenic nature of some of these changes (Drew 2014b, 35).

The analysis of Vijaya Rettakudi Nagarajan is also useful for conceptually mod-eling this complex situation because it offers an explanation for what might appear as a lack of consistency in human-nature relationships in South Asian contexts. In her theorizing of South Indian understandings of how the "Earth as Goddess

Bhu Devi" is understood to be present in *kolam* designs created by women as representations of auspiciousness and as a form of "painted prayer," Nagarajan (1998, 278) offers an explanation for why the religious importance of human-environment relationships has not been consistently upheld in the environmental history of South Asia. Drawing on the famous observations of A. K. Ramanujan about context-sensitive thinking, Nagarajan suggests that what she terms the "force" and "substance" of "sacrality" of the earth functions as a set of "embedded ecologies" that are present in daily life, but whose presence and reality are "intermittent" (278–79).[19] This idea helps Nagarajan explain why traditional ways of thinking that emphasize the closeness and interdependence of the human and the natural have led to the high levels of environmental degradation in India today: human perceptions of the sacredness of the earth inflect intermittently and in context-sensitive ways rather than continuously and pervasively.

In the months and now years after the floods of 2013 these ideas of intermittence, one-story-ness, and many thinged-ness help us understand the complexity of the emotional, moral, political, social, and ecological dilemmas faced by those tasked with rebuilding Kedarnath and returning there to try and make their living once again. In my view, however, one must keep in mind how these multiple variables at play in the Kedarnath place-situation functioned at the level of felt experience as a single thing, a single system characterized by the complex, nonlinear interplay of the forces of the market, the weather, the landscape, politics, trauma, *dosh*, and divine displeasure. Residents of the Himalaya were and are still feeling their way through the unfolding concatenation of phenomena relating to climate change, globalization, regional modernities, and an array of local, regional, and transregional religious worldviews. The causal relationships among these phenomena are highly variable and still, for many, opaque. Ehud Halperin's (2017) recent work on how phenomena relating to climate change interface with local religion in the Kullu valley of Himachal Pradesh captures this situation precisely.

What I want to emphasize at the close of this chapter is not so much how different meanings or events that can be found in the situation arose but more the felt, eco-social, emergent weight of this situation as a whole and how it affected the place at the beginning of the season in 2014. The enmeshed interplay of the presence and power of Shiva and Shakti, the multivalenced Himalayan *prakritic* location, the network of resident local divine powers, the ideas and practices of *vikas* produced by the political ecology of the region—the place gathered it all together as Kedarnath-Shiva continued to attract humans to its *daldali bhumi* (Hindi: marshy ground, the Hindi translation for the Sanskrit word *kedara*). Every time there was a period of intense rain or landslides, memories would come crashing back. This was the situation on the ground when I visited in 2014.

7

Topographies of Reinvention

Visiting the Kedarnath valley in 2014 was hard. Experiencing the grief and despair of residents in the area over and over again caused my emotions to numb out of self-protection. At the same time I felt great guilt—I was able to leave. This situation of constraint was not, at the end of the day, my situation. I was a scholar writing a book. A friend of mine in the Kedarnath valley told me that he had sent his family to live in Dehra Dun because he could not set aside his fear of landslides and floods but that he needed to stay because his business was based in the Kedarnath valley. Subjectively, I felt as if a weight lifted when I left the Kedarnath valley and was walking through the market in Srinagar. For the first time, I felt, in weeks, I saw groups of people smiling and laughing. In Dehra Dun it seemed like business as usual. I found a special issue of the English daily newspaper the *Garhwal Post* from the previous year entitled "Voices of Uttarakhand" in a bookshop just off Raipur Road, one of the most cosmopolitan streets in Uttarakhand. The dedicatory inscription read, "Garhwal Post pays homage to the thousands who lost their lives, property, and livelihood due to the combined impact of God's intervention and man's neglect." When I told the bookseller I was looking for *apda*-related material he said that he had almost nothing left from the previous year, when there had been a torrent. People have already forgotten, he said.

I visited Uttarakhand again briefly in January of 2017. Kedarnath was, of course, officially closed at the time. Reconstruction work, however, was ongoing. The Nehru Institute of Mountaineering (NIM) had workers and machinery in Kedarnath and was coordinating from its local office in Sonprayag in the Kedarnath valley. The institute, led by Colonel Ajay Kothiyal, was one of the most important players in the reconstruction of Kedarnath. Its involvement began almost immediately after the floods of 2013, when its mountaineering expertise became critically important for finding and rescuing survivors and mapping escape routes. Later, NIM took charge of many aspects of the reconstruction of Kedarnath: the construction of a new heavy-duty helipad for MI-26 cargo helicopters behind the temple, a three-layer semicircular boundary wall designed to protect Kedarnath against future flood events, and a new bathing *ghat* and bridge on the banks of the Mandakini at the southwestern edge of Kedarnath village. NIM also restored the path between Gaurikund and Kedarnath that, since 2013, has run along the eastern bank of the Mandakini after crossing the bridge near the ruins of what was once Rambara. It would be a massive understatement to observe that by January of 2017 this reconstruction campaign, accomplished in extremely challenging conditions, made Colonel Kotiyal, as the leader of these efforts by NIM, a public hero throughout Uttarakhand, renowned for both his competence and his commitment to the public good. NIM has also involved itself with another project relating to *yatra* in the Garhwali Dev Bhumi: the Nanda Devi Royal Yatra (Hindi: *Nanda Devi Raj Jat,* once famously described by William Sax).

Checking in with friends and acquaintances during this visit was sometimes uncomfortable. Talking with me brought back their memories and experiences of 2013 that, I think, people were trying to leave behind as best they could—trying to move past grief over lost loved ones and the fraught memory-fear that bad weather now evoked. I had not spent enough recent time in the Kedarnath valley to feel fully caught up with relationships and with what was going on. As we navigated these tensions I found, among the friends and acquaintances with whom I spoke, a cautious optimism. The *yatri* numbers were increasing again. New systems were in place—biometric registration of *yatris,* increased government attention to road and weather conditions, the undiminished involvement of the state government. Friends of mine who had two years before made the risky decision to wait out a lean period and remain connected to *yatra* tourism in Kedarnath were starting to feel a bit relieved. At the same time, many earlier patterns were still unfolding but with new iterations. There continued to be widespread chai-shop controversy on the kinds of relief provided to Kedarnath survivors and the families of the dead. The ability of Kedarnath locals who owned property in and around Kedarnath itself to rebuild their businesses continued to be in conflict with how NIM, the Samiti, the Rudraprayag district (Hindi: *zila*) government, and the state government were approaching reconstruction in Kedarnath. The number of *yatris* using helicopter services seemed, anecdotally, to be on the rise.

The recent past was becoming part of the longer story of the place. The journey to Kedarnath now layers in legends of disaster. The path from Gaurikund, after crossing over to the eastern side of the Mandakini near where Rambara used to be, carries the memory of where the old path used to run. Particularly above Linchauli, the new midpoint, the new path is more difficult than the old path, and

I suspect we will now see that when family groups visit Kedarnath on the Char Dham some but not all will make it all the way to Kedarnath. GMVN's free food and lodgings for *yatris* may become a permanent part of the place. Here is where Rambara used to be. Here is the *divya shila* that protected the temple. Behind the village is where the cargo helicopters can now land, and behind that is the boundary wall. The built environment of Kedarnath now models a particular relationship, almost a challenge, to the natural tendencies of the place—a claim that what humans have built after the floods can withstand whatever comes next. Implicit in this built environment is a redoubled commitment to the power of government-supported civil engineering and environmental science.

SENSING ECO-SOCIAL CONNECTIONS

This book orbits the idea that it is useful to view experiences of Kedarnath as experiences of a place that functions as an eco-social system characterized by complexity. In this place Shiva, the Goddess, human actors, cultural forms, objects, economic networks, the landscape, and the weather are all variables that interact with each other and that are in certain ways almost multiforms of each other. This current state of affairs is the result of the partially continuous and partially disjunctive, emergent blending of premodern patterns of Himalayan, Shaiva, and Shakta eco-social thought and practice with modern, colonial, and postcolonial relationships to Hindu traditions, the natural world, religious travel, tourism, and development. The roughly chronological story I have told shows how the pursuit of *dharma* carried out by the Pandavas came to be woven together with first-millennium CE understandings of Shiva's presence in the Himalaya in a way that constructed Kedarnath as a place of purification, violence, wounding, blessing, transformation, presence, and absence. Kedarnath became a somewhat inaccessible node on the subcontinent-wide network of sacred geography attested in Puranic and Upapuranic texts. It became part of the transregional webs of different Shaiva groups. It became a destination for kings and renunciants, a point of intersection between the "below" of the Gangetic plain and the higher worlds of the Himalayan hills, mountains, and glaciers. In the last two centuries it became a destination for people, mostly Hindus, from all walks of life. The Kedarnath valley became part of British Garhwal, and Kedarnath, along with Badrinath, came to be a point of contention between the British administrators of British Garhwal and the kings of Tehri. The Kedarnath valley was important during the Chipko movement and, like much of the Garhwal Himalaya, is part of a region that has been marked for well over a century by contestation over land use (particularly tree cutting) and the control of water (the canalization of the Ganga in Haridwar, one of the gateways to Garhwal, and the construction of the Tehri Dam and other large-scale hydroelectric projects). These projects were based on an instrumental relationship to the natural world that construed woods and water as resources to be mined and controlled for the benefit of those outside the region. Travel to Kedarnath and Badrinath came to be done as part of the Char Dham Yatra to Yamunotri, Gangotri, Kedarnath, and Badrinath. This pilgrimage, along with Sikh pilgrimage to Hemkund Sahib, fed into the construction of the region as Dev Bhumi, the Land of the Gods, a modern commercial appellation based on multiple layers of premodern reverence for this section of the Himalaya, whose shrines are particularly well attested in Puranic literature. As *yatra* gave way to *yatra* tourism, the Himalaya and all the divine powers resident there became connected to modern Western ideas about "nature" and natural beauty that were based partially on the idea that humans were somehow separate from nature. Specific understandings of Himalayan nature became formative for regional identity and became part of the increasingly touristic and nature-oriented practices and expectations of visitors coming for the Char Dham Yatra. The rise in the number of middle-class visitors in recent decades combined with the new state's strong but short-sighted support for *yatra* tourism to produce a situation where traditional knowledge and practice about how to live, build, and travel in the Himalaya was overlooked. As a result, an instrumental relationship to the natural world was enacted on a hitherto unprecedented scale. Roads were widened. Hotels were built on floodplains. Visitor numbers were not regulated. The floods of 2013 punctured and exposed this rapidly expanding, unsustainable growth. Because of how premodern stories, experiences, understandings, and practices connected Shiva, Devi, *shakti* (and thereby *prakriti*), and myriad *devtas* with the land and water of *kedaramandala*, an intense, complex, fluid sense of divine presence and power is woven throughout this emplaced situation. This wovenness, this *enmeshing,* characterizes the postflood situation in ways that are sometimes opaque and sometimes transparent even as the situation continues to unfold and change as we move away in time from the floods. Kedarnath, by October 1 of the 2017 pilgrimage season, had reportedly welcomed 441,770 visitors (Shree Badarinath-Shri Kedarnath Temples Committee 2017).

I have written this account to give a sense of both the feeling and the materiality of this powerful place, the profoundly interconnected nature of these different arenas and contexts, and the way that this "system" is situated in space and time. I chose to write in this way because this is what I think a scholar of religion has to contribute to our understanding of the situation. My training and voice are not necessary to expose the profound ways that failures of the (postcolonial) state intensified the impact of the flooding and helped create the conditions leading up to that intensified impact. These are matters of public record. So why this book? First, it is a way of bearing witness, and bearing witness is an act that may resound in unpredictable ways.

Second, my experience, research, and training have allowed me to see a depth of connection in this situation between what are often construed as separate arenas

FIGURE 15. Kedarnath in 2017. Photograph by Madelyn P. Ramachandran.

and contexts. It is my hope that social scientists, natural scientists, policy makers, and elites of diverse types will make better decisions, construct better research agendas, and be more willing to explore new forms of collaboration after encountering robust and portable ways of thinking about the embeddedness of "religion" in human life. By "robust and portable ways" I mean models that show how phenomena connected to what some people might sometimes term "religion" are pervasively embedded across and within the myriad contexts of human life that, as Anna Tsing (2015, 3) has observed in her work on the rainforests of Indonesia, are bound together through "the productive friction of global connections." In "basic empirical research" based on interviews at the 2012 United Nations Conference on Sustainable Development (Rio+20), Evan Berry (2014, 269) reported that "content analysis of interview responses suggests that religious actors hold divergent views about the salience of religion to global sustainability politics. The central finding is that the boundary between religious and secular civil groups is a permeable one." It is more difficult, however, to conceptualize exactly how this (contingent) boundary is permeable and under what circumstances. This is an example of an area where I think attention to Kedarnath is beneficial. Close attention to Kedarnath acts as a focusing lens that allows us to see how places and regions "gather" together (recalling the discussion of Edward Casey in the Introduction) matters of religious, ecological, political, cultural, and economic concern. There is an important implication here. Development processes on the ground may not be wholly "secular," and attention to the "religious" aspects of how people understand place and environment illustrates this. Politicians in Uttarakhand are responsible to a diverse set of publics that include divine agents and the Himalaya it/themselves (on this point in the Himachal context, see Berti 2009). This way of thinking about Kedarnath does not require a commitment to the idea of religion or religious experience as *sui generis*—that is, as transcendentally given categories that constitute wholly distinct spheres of human experience and activity.

Yet at the same time I make a gentle assertion here that the category of "religion" has a *heuristic* utility. The set of human actors, phenomena, ideas, processes, and materialities that are often denoted by this term powerfully shape the world. The remarks of Peter van der Veer (2014) in his recent comparison of the relationships of religion to the state in India and China are instructive in this regard. Van der Veer notes that "world history more often than not emphasizes economic and politics and in an established secularist fashion underplays the formative role of religion" (9). As a way of moving past this set of analytic limits, he proposes "an emphasis on what I call a 'syntagmatic chain of religion-magic-secularity-spirituality.' I borrow the term 'syntagmatic' from Saussurean linguistics and use it in a nonlinguistic manner to suggest that these terms are connected, belong to each other, but cannot replace each other. They do not possess stable meanings independently from one another and thus cannot be simply defined separately. They emerge historically together, imply one another, and function as nodes within a shifting field of power. This syntagmatic chain occupies a key position in nationalist imaginings of modernity" (9). As we have seen in the case of Kedarnath and more broadly Uttarakhand, these "nationalist imaginings of modernity" (as well as their more regional iterations) impinge directly and with increasing magnitude on the conflicting expectations placed on both state and center for the regulation of pilgrimage tourism, the protection of the environment, the stimulation of local and regional economies, and the construction and maintenance of civil and commercial infrastructure connected to pilgrimage tourism. Prime Minister Narendra Modi visited Kedarnath in May of 2017. The director ("principal") of NIM is typically a highly decorated officer from a branch of the Indian military who is chosen by the Ministry of Defense.

Yet—and I think this is part of why Van der Veer's argument is important—the discursive frame of "nationalist imaginings" is itself not, without further nuance, well prepared to address a situation in which the Himalaya are both divine parent and father-in-law, the world's youngest and most active mountain range, the water table for much of Asia, the residence of countless place-based deities whose efficacious actions leave traces in our world, the residence and at the same time multiform of Shiva and Shakti, and the host for myriad micro- and regional economies whose parameters are in important ways set by the terrain and the weather, which now themselves bear the mark of human agencies. Holistic approaches have a role to play in understanding this kind of situation because they help find a way past binaries such as human/nature, secular/religious, or science/religion that limit the crafting of sustainable policies that make ecological sense. This book may be of assistance in communicating a sense of what Adrian J. Ivakhiv (2012, 226), building on the work of Isabelle Stengers, Bruno Latour, and Philippe Descola, wants to call the increasingly "cosmopolitical" way that religion, science, nature, and politics (all at multiple scales) are dynamically interdependent, and the way that the world may be moving toward this realization.

In Himalayan contexts, scholars have already been organizing in transregional and multidisciplinary groups to produce precisely these kinds of analyses so that the results can achieve breadth, depth, and at the same time utility and relevance for those whose lives are directly affected. This way of working has become all the more important after not only the floods of 2013 in Uttarakhand but also the major earthquake of 2015 in Nepal whose epicenter lay northwest of Kathmandu. The Everyday Religion and Sustainable Environments in the Himalaya initiative of the New School's India-China Institute (and their follow-up "Sacred Himalaya" initiative) make precisely this point. They state their overarching goals in this way: "The project aimed to create an enabling environment for knowledge-sharing and production on the complex role of religion with particular emphasis on sustainable environmental issues" (India-China Institute n.d.). Mabel Denzin Gergan

(2017, 490; 2015), as part of a broader postcolonial theoretical invitation to geographers to think beyond "modern secular tendencies," has shown "how people's relationship with a sacred, animate landscape is not easily translatable into the clear goals of environmental politics." Daniela Berti (2016, 2015) and Debjani Bhattacharyya (2017) have demonstrated a similarly complex set of relations that unfold when divine agents and natural phenomena are potentially viewed as legally significant "persons" in the Indian legal system.

The academic study of religion can offer holistic ways of thinking about such situations that both draw on the deep wells of knowledge and understanding about the natural world found in what some call "religious traditions" and also, by putting specific examples into a more general theoretical conversation, hold partly at a distance the authority and influence of religious actors themselves. These new second-order holisms, approached through ideas of system, flow, place, experience, and network, can offer support for the kinds of interdisciplinary, global, collaborative projects humans need to survive and flourish in the twenty-first century. This book has been an attempt to model the work involved in constructing a bridge from the particular to the general in a way that neglects neither and that models a holistic way of approaching the interrelationships of "religion," earth, and human experiences of being-in-the-world. The idea of eco-social complexity, an idea that can be understood as a specific form of this type of holism, is a useful way of seeing how the "syntagmatic chain of religion-magic-secularity-spirituality" connects over time to climate destabilization, environmental change, disaster management, religious tourism, global flows of capital, and competing visions of development. This way of thinking provides a portable metaframework for thinking about ecology and development that seamlessly includes religion. It has two advantages—it models using some tools that natural scientists and social scientists are already familiar with, and it moves past the limiting, inherited, often colonial Western binaries of human/nature, religious/secular, and modern/premodern ("traditional"). It also offers usable application parameters: the analytic category of *place*.

Even though I am clearly insisting that the religious dimensions of place and region be further integrated into broader approaches to questions of policy matters such as sustainable development, this methodological invitation should not be taken as a recommendation that endorses the idea of a Hindu India against the political projects of secularism.[1] It should rather be taken as a descriptive, pragmatic assessment of the fact that the religious resonances of reconstruction in and around Kedarnath are very much bound up with the political, economic, and environmental considerations that can be used to construct durable, beneficial frameworks. These are not fully separable conversations, and the political implications of this fact are difficult to predict. Daniela Berti (2015, 113) made precisely this point when, in her article "Gods' Rights vs Hydroelectric Projects: Environmental Conflicts and the Judicialization of Nature in India," she observed that "this mutual association between religion and ecology does not always take a Hindutva turn." State politicians in Uttarakhand are in a sense responsible to a constituency that includes local deities, local forms of deities and emplaced forms of holiness and divine power who are known beyond the region, and the polyvalent massive *tattva* (suchness) of the Himalaya and rivers such as the Ganga and Yamuna.[2] Mark Elmore (2016) has shown how, in the neighboring *pahari* state of Himachal Pradesh, the very idea of a Himachali "religious" regional identity emerges out of and is in complex interaction with imagined secular modernisms.

Where does this leave us? How are we to view the situation in the Kedarnath valley, and more broadly in Uttarakhand, today? What can be learned? Part of what can be learned is that the complex, overlapping, intersecting relationships among divine powers, natural phenomena, and humans will continue to be a part of whatever shape human life in this region takes. We may also observe that there is something approaching a consensus that Uttarakhand needs to move in more sustainable directions when it can do so in ways that are politically and economically practical in the short and middle term. This can be seen in discussions and recommendations about different forms of pilgrimage and tourism, hydropower, forest management, agriculture, and employment (Chopra 2014). With regard to tourism specifically, in its diverse and changing forms (some of which include *yatra*), public sentiment in Uttarakhand includes the idea that the state needs to create conditions that favor visitor activity that is more decentralized, more attentive to the environmental cost of human activity in Himalayan landscapes, more concerned with local and regional culture, better regulated, and spread throughout the year. Nature and trekking-oriented tourism, cultural tourism, and ecotourism may increase.

However, it would be a mistake to imagine that this new "Bhumi" would be less "Dev." As I began by saying in the Introduction, one of the main goals of this book is to create a set of feelings, of interpretive attitudes, that will assist readers in thinking and feeling *holistically* about the connections of religion, ecology, development, and disaster, particularly in the Himalaya. This is consonant with the fact that much of my approach is inspired by the tradition of phenomenological anthropology. The complexity of the situation does not reduce to the ecological any more than it does to the political or the economic or the religious. This cautionary note has been sounded before by scholars who have written about Hinduism and ecology (Tomalin 2004; Haberman 2013, 195). It is not yet clear to me whether the Kedarnath situation can be held up as an example of how nature-focused religious sentiments create ecological resilience, a function that several scholars have argued is provided in South Asian contexts by sacred groves (Kent 2013; Gadgil and Guha 1995, 185). Rather, constructions and experiences at Kedarnath of place and god render transparent the complex intertwining of forces that are ubiquitous and forces whose persistent patterns as a whole exceed the sum of their parts—an example of what Philip Fountain and Levi McLaughlin (2016, 2) termed, in the context of a guest editor's introduction to a special issue of *Asian Ethnology* on religion, disaster relief, and reconstruction, "religion *in situ*." The situation does not reduce to how religious environmental groups can influence policy, how indigenous groups have begun to achieve collective political power in solidarity, or how landscape changes create the conditions for shifts in political advantage. It does not reduce to neo-Gandhian approaches to sustainable development or to the intensely totalizing, recursive, self-transformative world of Shiva *bhakti*. It does not reduce to how nation-states in the Global South approach questions of sustainable development. It may or may not be a case study that provides encouraging resources for doing what John Grim and Mary Evelyn Tucker (2014) have famously termed "religious ecology," the collaborative investigation of the ecologically productive aspects of lived religious worldviews. Rather, attention to Kedarnath offers a model for what Ivakhiv, during a panel presentation at a weeklong seminar organized by the School for Advanced Research on the Human Experience in Santa Fe, termed a "more complex understanding" (Tucker 2012, 6) of "human relationships with their socionatural and built environments."[3] I think that complex understandings of this sort, if sufficiently engaged, facilitate modes of action that can be beneficial for our planetary community.

OF BOUNDARIES AND RESILIENCE

The new three-tier boundary wall constructed behind Kedarnath can, stand as an icon that signals the necessity of approaches based on these sorts of complex understandings. I mused previously that the construction of the boundary wall to protect Kedarnath village could be viewed as a redoubled challenge that asserted the ability of humans to shape the natural environment to their own needs. But the construction of this boundary wall bears closer analysis because it stands as a material testament to the complex character of nature-landscape-human relationships. The wall has three layers. The purpose of the first layer (closest to the glaciers, farthest from the village) is to divert the flows of the Mandakini and Saraswati rivers around Kedarnath while at the same time partially arresting the movement of debris in the water in a way that helps the debris to become part of the wall, thereby strengthening it. The second wall is designed to catch the overflow from the first wall, and it too is designed to arrest debris. Then, closest to the village, a third wall stands ready to catch the overflow from the second.[4] In other words, the system assumes that it cannot simply stop the flow of floodwater and debris but rather must direct, filter, and join with the aquatic power that might again, someday, descend on the village. One might almost say that it is a human-produced microform of how, in the descent of the Ganga, Shiva's hair filters her otherwise unbearable power—a twenty-first-century example of *sarupata* in which the wall makers become like the deity. Is this an example of eco-socially aware infrastructure, a piece of the built environment that appropriately acknowledges how resilient humans might fit into this Himalayan situation in a way that honors older understandings about human-river-divine relationships in the Himalaya? Or is it an encouragement to humans in Kedarnath to feel safer than they ought to feel? Could it be both? Does this wall foster hope? If it does, is it sustainable hope?

SUSTAINABILITY AND HOPE

The possibility of hope leads us to a final question of critical global import that my study of Kedarnath allows us to consider: how the idea of sustainability connects to matters of religious importance and how it can be shaped or informed by religious perspectives. After tracing the specific history through which the ideas of "sustainability" and "development" came to be coupled, environmental ethicist Willis Jenkins (2011, 108) prophetically argued that questions of sustainability in the twenty-first century, questions that will in one way or another continue to occupy Uttarakhand for a long time, should proceed in conversation with religious resources if the specific resources in question appear to have the capacity to "sustain hope" for humanity. Jenkins notes that we need a framework that offers criteria for deciding what humans ought to sustain, and that religious frameworks have a great deal of expertise in addressing the existential and moral challenges raised by such a question: "Making sense of sustainability requires at least recognizing the religious responses and questions that this challenge to humanity provokes" (108). My suggestion is that close attention to places of religious significance can teach us a good deal in this regard. More specifically, close attention to places of religious significance can teach us about the material, experiential aspects of sustainable hope. A brief consideration of an excerpt from a July 2013 interview given by the famous environmental activist Sunderlal Bahuguna will show what I mean:

> Interviewer: A lot of the damage inflicted is also by locals. Do you think the people of Uttarakhand generally are still concerned about their environment?
> Bahuguna: Yes, definitely. There have been so many other movements after Chipko that have been pushing for a local resource-based economy, protecting eco-sensitive zones and our rivers. When has the voice of the rural people ever been heard? The government always claimed their cause was an emotional one but they can't say so after this disaster. This is a lesson and we must change our policies. (Bahuguna 2013)

The term *local resource-based economy* fits in neatly with many of the economic and political discursive frameworks connected to the ideas of sustainable development that undergird many current conversations about the future directions necessary for Uttarakhand. Terms like this are part of the lexica of environmental studies and geography. But this particular term also extends into Van der Veer's "'syntagmatic chain of religion-magic-secularity-spirituality.'" Bahuguna's exhortation is grounded in his broader sense of a nature-spirituality that is in important ways inspired by the teachings of Gandhi and the wisdom found in the *Bhagavad Gita* and various *Upanishads* (James 2013, 205–25). It is, therefore, reasonable to read Bahuguna's call for more serious government commitment to sustainable policies as a sentiment produced by both a grassroots egalitarian dedication to social and environmental justice and a deep sense of the sacred and interconnected nature of life on earth, which makes it a good example of the kind of "religious response" Jenkins wants to engage.

I want to suggest how Bahuguna's remarks about sustainability might apply to the specific terrain of the Kedarnath valley and might be received by the people who live or visit there. I think that the ambit of the term *local resource-based economy*, viewed from the perspective I have developed in this work, engages the entire material-cultural-political-economic-environmental matrix signified by the term *prakriti* discussed in chapter 6. It includes not just trees but the deities who reside in and own forest areas and who have shrines on mountaintops—deities who themselves occasionally travel on *yatra* along with human devotees from all over the world to be recharged by Shiva in his Kedarnath form, in that place where the Kedarnath *linga* of light emerges from the ground near the source of the Mandakini River. It also includes all the connections (ecological, social, political, economic) among and across divine agents, humans, and the landscape that connects them. I think that this emplaced sense of interconnectedness, what I have been calling the experience of ecosociality in the complex Kedarnath place-system, itself exerts in the experience of humans present in Kedarnath a kind of material weight that is similar to the influence of a physical part of the landscape, such as the Kedarnath boundary wall or the weather experienced on the path. The sheer physicality, the overall somatic load, of the experience of ecosociality in places like Kedarnath should somehow be part of conversations about sustainable forms of hope. This is my critical phenomenological extension of the suggestion of Manuel Vásquez (2011), discussed in the Introduction, that a focus on materiality can lead the study of religion in an ecological direction. The ecosociality of the place is *itself* sensed and felt. It is material. Part of what sustains hope in Uttarakhand, and more broadly in India and South Asia, is the material fact of an ongoing, emergent, dense, thick connectedness to systems and networks of efficacious divine power present in the natural environment—systems and networks with which humans can enter into relation in both day-to-day ways (in the case of residents) and occasional ways (in the case of visitors).

From my conversations with many locals and visitors in Kedarnath valley I got the sense of this kind of thick connectedness as something ubiquitous at the experiential level even though seldom expressed directly; and it is an important, though difficult to articulate, reason that specific people might choose to engage in some form of environmental activism or choose to act in ecologically beneficial ways. The material fact of connectedness is different from, but a necessary ingredient of, the ideas, concepts, discourses, ideologies, and beliefs, ecologically positive or negative, that emerge from it. It is part of the stuff of our world that can, under the right circumstances, animate movements toward sustainable forms of hope. If it can be termed a form of spirituality or religion, then it is an intensely material spirituality or religion. Scholars of place and pilgrimage in South Asia are to some extent already attentive to the experience of this connectedness (see, for example, Eck 2012: Feldhaus 2003; Sarbadhikary 2015). I want to emphasize the *prakritic* weight of this connectedness and show how it might affect how people experience and engage ecological issues, and how understanding of such matters can be achieved through close attention to specific places.

Kedarnath is a place where, with close attention, it is possible to acquire a sense of the energy and power of this material ecosocial connectedness—to feel where it runs especially close to the surface, how it pulls you toward it, and how it contributes to human experiences of being in the world. I hope that we can then take this way of attending to places of religious importance and apply it to other cases, thereby deepening our collective ecological understanding of how humans fit into the world in ways that can hopefully be leveraged toward a (relatively) hopeful future. This is where I leave the reader: with a portable sense of both the experiential weight of these systems of religious eco-social connection and an understanding of how this complex experiential weight is produced through place. *Jai Kedar.*

GLOSSARY

abhishek	ritual bathing of a deity/sacred object with auspicious materials
acheri	Garhwali forest/mountain spirit
Agama	a set of Shiva-focused ritual and philosophical texts
apda	disaster
arati	evening worship offering light and songs to deities
Badrinath	Himalayan shrine of the god Vishnu, part of the Uttarakhand Char Dham Yatra
Bhagwan	God, Lord
Bhairav(a), Bhairavnath	a fierce deity who became known as Shiva's lieutenant
bhajan	devotional song
bhog	food offerings to a deity
Bhukund Bhairavnath	the form of Bhairavnath resident in Kedarnath
bhumi	earth, land
BJP	Bharatiya Janata Party
Chorabari Tal	(tal = Hindi: lake) the snowmelt lake above Kedarnath that descended in the Himalayan Tsunami, also known as Gandhi Sarovar
dakshina	ritual fees

Dashanami	a monastic order of Shaiva renunciants
deora	*yatra* of a Garhwali deity
Dev Bhumi	Land of the Gods, state-promoted name for the Garwhal region
devdarshini	a vantage point from which you are first able to glimpse the deity of a particular place
Devi	or Mahadevi, the Goddess
devi	a form of the Goddess
devra	same as *deora*
devta	a deity
dham	an abode or dwelling place of a deity
dharamshala	pilgrim guesthouse
divya shila	"divine rock," the massive boulder that stopped in front of the temple in the 2013 floods and shielded the temple from further debris flow
doli	Hindi, Garhwali: palanquin
Durga	warrior goddess, a manifestation of the Goddess
Durga Puja	annual Hindu festival for the goddess Durga
gana(s)	the fierce followers or troops of Shiva; manifestations of Shiva's anger
Gandhi Sarovar	another name for Chorabari Tal
Gangotri	shrine of the goddess/river Ganga, part of the Uttarakhand Char Dham Yatra
Garwhal	Himalayan region in which Kedarnath is located, today part of the state of Uttarakhand
Garhwal Mandal Vikas Nigam	Garhwal Regional Development Authority
Gaurikund	village where the footpath to Kedarnath begins
ghat	steps to river for bathing, bathing area with steps leading down to it
GMVN	Garhwal Mandal Vikas Nigam
havan	ritual of fire offering
ishtadevta	personal deity
jyotirlinga (Sanskrit) / *jyotirling* (Hindi)	a *linga* of light, a vibrating column of light that manifests Shiva's universal form; one of twelve locations spread throughout India where such a column of light embeds itself in the ground

Kalamukhas	a Shiva-oriented group that focused on using transgressive behavior as a way to move past mundane distinctions of pure and impure
Kali	a manifestation of the anger of the Goddess
Kapalikas	a sect of those who worship Shiva in his Bhairava form
katha	recitations of famous stories from epics and *Puranas*
kedaramandala	word for Kedara region in Sanskrit texts
Kedareshvara	synonym for Kedarnath-Shiva in Sanskrit
Kedarnath	a pilgrimage site that is an abode of Shiva, and the village located there. Literally, "Lord of the Marshy Ground"
Kumaon	neighboring region to Garwhal, today part of the state of Uttarakhand
langar	free food kitchen
linga	the cylindrical shaft that is the aniconic form of Shiva's presence in the world
Mahadevi	the "Great Goddess" of Hindu traditions
Mahapath/Mahapanth	the Great Path; the path that the Pandavas took in the Climb to Heaven
Mahaprasthanika	The chapter of the Great Departure, one of the chapters of the Sanskrit *Mahabharata* connected to Kedarnath
Maheshvara	Great Lord (another name for Shiva)
mantra	powerful sacred verse
Mantramarga	the path of mantras, an early Shaiva tradition
math	headquarters
mela	festival
murti	Hindi: material form of a deity
-nath	lord
Naths, Nath-Siddhas	a tantric Shaiva sect partially descended from the Kalamukhas with a substantial presence in Garhwal and Kumaon
naur	Garhwali term for a designated human representative of the deity, often one whom the deity regularly possesses
Panch Kedar	group of five Shaiva temples in central Garhwal that includes Kedarnath, Madmaheswar, Tungnath, Kalpanath, and Rudranath

pap	wickedness (Hindi)
papa	wickedness (Sanskrit)
parikrama	Hindi: circumambulation
Pashupatas	early Shaiva group
Pashupati	Lord of Cattle/Beasts, often understood as an epithet of Shiva
prakriti	Hindi: nature, also a form of the Goddess in some contexts
prasad	devotional offerings shared with God, typically food
pravacan	teaching sermon
puja	ritual worship that contains a sense of offering and reverence
pujari	Hindi: ritual specialist tasked with the daily worship of temple deities
Rudraprayag	the district in which Kedarnath is located; also the name of a town within the district
saddhaka	*siddha*, tantric practitioner
sakshat darshan	a visual encounter with (or visual knowing of) the divine in its most true form (*sakshat* = actual)
Samiti	(committee, assembly); Badrinath-Kedarnath Temples Committee
sarupata	the state of having the same form as a specific deity
Shaiva	Shiva-oriented
Shaiva Siddhanta	an early Shaiva system of devotional practice and thought
shakti	force, power, energy, authority; personified, as a specific goddess, who is both an independent being and the consort of a god such as Vishnu or Shiva
Shakti	one form of Mahadevi, the "Great Goddess" of Hindu traditions
Shankara	famous Hindu philosopher who is believed by many to have visited the central Indian Himalaya before the end of his life
shraddh	Hindi: the rituals of ancestor worship
shraddha	faith
siddhi	supernatural powers

svayambhu	a self-manifested form of the deity (as opposed to one created by human beings)
swargarohan	in Hindi, the Climb to Heaven made by the Pandava family
svarupa (Sanskrit) / *svarup* (Hindi)	true form (e.g., of a god) (*rupa / rup* = form)
tirtha	a place that offers the possibility that one can "cross over" or "ford" the ocean of rebirth
tirth purohit	pilgrimage priest
Udak Kund	the Udaka water tank in Kedarnath
Ukhimath	the town where Kedarnath-Shiva is worshipped in the winter months when Kedarnath is closed
Uttarakhand	state in which the Garwhal region and the Kedarnath site are located
Uttarakhand Char Dham Yatra	Hindi: Uttarakhand Four Abode Pilgrimage. A journey to the abodes of four deities: Yamunotri, Gangotri, Kedarnath, and Badrinath
Ved-pathi	Veda-reciter employed by the Badrinath-Kedarnath Temple to assist devotees in performing *puja* in the Kedarnath temple
Virashaiva	a Shiva-oriented denomination; members are Virashaivas. This term has a complicated relationship with another Shaiva group known as Lingayats.
Yamunotri	shrine of the goddess/river Yamuna, part of the Uttarakhand Char Dham Yatra
yatra	a term of broad and fuzzy semantic range that can mean pilgrimage, tour, or vacation

NOTES

INTRODUCTION

1. Personal communication, June 1, 2007, Kedarnath.

2. Richard Davis (1995, 639) has usefully offered a more involved definition: "The word *liṅga* has three primary meanings, and all three are important here. *Liṅga* denotes the penis, the male generative organ. It also denotes a mark, emblem, badge—a sign that allows one to identify or recognize something, as one may identify someone as a member of the male sex by his penis. Finally, it also denotes the primary cult object of Śaivism, an upraised cylindrical shaft with rounded top, rising from a rounded base. The icon resembles, in a generally abstract manner, an erect male member, and serves at the same time as a sign of Śiva."

3. As Don Handelman and David Shulman (2004, 37–38) put it: "The *liṅga*, in short, is the core of this god at home in our world: an ontology of pure potentiality is present here." The Kedarnath situation is further complicated by the fact that not everyone experiences Shiva's Kedarnath form as a *linga*. I am grateful to Aftab Jassal for suggesting I emphasize the remarks of Handelman and Shulman on this point.

4. While the *jyotirlingas* are commonly thought of as a set of twelve, the names and locations in this set vary. I discuss this in more detail in chapter 3. For an overview of the *jyotirlingas*, see Fleming (2009b).

5. I am grateful to the anonymous reviewer from the University of California Press for the suggestion, here and throughout the manuscript, that the precarity of visiting and residing in the Himalaya deserves emphasis.

6. Personal communication, June 23, 2007, Kedarnath.

7. I employ pseudonyms when describing these kinds of conversations.

8. "The attraction [ākarṣan] of the Himalaya is nowhere else" (Ḍabarāl, n.d., 5). The published copy of this work available to me did not contain a publication date. It is, however,

reasonable to assume that this work was published in approximately the early 1960s. It does not discuss in detail additional road building in the region brought on by the Sino-Indian Border Conflict of 1962, and it notes that the motor road to Kedarnath extends through Guptkashi (254). It refers in several places (e.g., 585) to 1951 Census of India data, which suggests that 1961 census data were not yet publicly available in published form. Ḍabarāl's description of *yatra* conditions demonstrates rough concord with that given in the famous pilgrimage digest *Kalyāṇ Tīrthāṅk*, which was first published in 1957 (Podār, Goswāmī, and Shāstri 1957). I am grateful to James Lochtefeld for his assistance in dating Ḍabarāl.

9. Personal communication, May 28, 2007.

10. Personal communication, May 28, 2007.

11. I am indebted to Laurie Patton and Joyce Flueckiger for this notion of "residue."

12. "Śiv-Jī aur Kedārnāth vahī cīz haiṁ" (personal communication, May 2007).

13. For a full discussion of commercial visual culture in Kedarnath, see my previously published article in *Material Religion: The Journal of Objects, Art, and Belief:* (Whitmore 2012).

14. There are, today, many different ways to understand the relationship between "place" and its analytic partner "space." Some regard space as the parent category, and some do not. See Knott (2009); Kong (2010); Gaenszle and Gengnagel (2006, 8); Lochtefeld (2010, 4–5); Feldhaus (2003, 5); Hubbard, Kitchin, and Valentine (2004); Warf and Arias (2009); Ferguson and Gupta (1997); Low and Lawrence-Zunigais (2003); Feld and Basso (1996, 4–11). In the study of religion, J. Z. Smith's (1987) illustration of how places constitute and reflect the projects of ritual has been formative for establishing the value of place as a category of analysis.

15. For an earlier application of the idea of complexity to place and pilgrimage in South Asia, see the edited volume *Pilgrimage: Sacred Landscapes and Self-Organized Complexity* (Malville and Saraswati 2009).

16. The question of how to think about scale is central to human geography and political ecology (Neumann 2009). As Francisca Cho and Richard Squier (2013, 373) note, "In thinking about a complex system, we have to balance our choice of scale against our scope of interest."

17. My understanding of place as complex agent draws on a preexisting theoretical trend in the study of South Asia to understand deities as complex agents (for this theoretical genealogy, see Sax 2009, 94–95; Latour 2005, 1993; Inden 1990).

18. For an extended consideration of the limitations of this binary, see Descola (2013). See also Bauer and Bhan (2018); Gergan (2017); and Chakraborty (2018).

19. For recent, substantive examples of this trend, see Tweed (2008) and Vásquez (2011).

1. IN PURSUIT OF SHIVA

1. I am grateful to Andrea Pinkney for the idea of Dev Bhumi as a brand.

2. See Elmore (2016, 115). I am grateful to Udi Halperin for his guidance on this point.

3. A common story in the area is told of Banasur's daughter, Usha, who dreamed of a man she did not know and fell in love with him. She related her dream to her friend, Chitralekha ("picture-writer"), who had the ability to create portraits based on verbal

descriptions. It was revealed that Usha had fallen in love with Aniruddh, the grandson of Krishna. Usha kidnapped Aniruddh and married him. Krishna went to war with Banasur, and Shiva was forced to make peace. In Ukhimath, the far and away most ancient part of the temple complex (judging by how much lower the floor is in that part of the complex) is the marriage place (Hindi: *vivah-sthal*) of Usha and Aniruddh. In the village of Lamgaundi, in the Bamsu area that belongs to Banasur, there is a small shrine where the image of Aniruddh is worshipped. If you are arriving from Guptkashi you enter this area by crossing the Ravan Ganga River. Today in Ukhimath residents of the village of Dangwari feel themselves to be particularly connected to this history. Dangwari has a tradition of producing artists and artisans, a tradition often ascribed to the arrival of South Indian artisans (Hindi: *shilpakar*) who were somehow connected to the Virashaivas. The largest photographic studio in Ukhimath is called Chitralekha.

4. *Tirth purohit* accounts of this history often state that centuries ago most of the important shrines in the Kedarnath valley were originally under the control of the 360 original families. Some Kedarnath *tirth purohits* have, since 2013, established a website that details their concerns and responses to the disaster. See "360 Teerth Purohit" (n.d.).

5. James Lochtefeld, personal communication, April, 2010.

6. For a description of the Ramanandis, a renunciant organization that includes both monastics and householders and that, as Peter van der Veer has argued, demonstrates an extremely fluid and open complex of community identity and practice, see van der Veer's *Gods on Earth* (1997).

7. See Shobhi (2005, 286) for a discussion of how the meaning of the term *jangama* in Virashaiva and Lingayat contexts changes from "wandering ascetic" to "caste group."

8. The actual term used here was "back part" (Hindi: *pṛṣṭh-bhāg*) of the buffalo. This detail refers to a part of the story that, in Tiwari-Ji's telling, we have not yet reached and that will be made clear later in the chapter.

9. I am grateful to Vidhu Shekhar Chaturvedi of the American Institute of Indian Studies for this semantic insight.

10. The art historian B.N. Goswamy (2013) has connected this text to a series of paintings (ca. 1815) in the *pahari* (mountain) style by artists working in the nearby region of Kangra, located in what is today known as Himachal Pradesh, created around 1815. I am grateful to both Ronald Davidson and Shaman Hatley for this reference.

11. I am grateful for the guidance of Fred Smith in thinking about this point.

12. I am grateful to Abigail Sone for her input on this point.

13. I am grateful to D.R. Purohit for his invaluable assistance in translating and understanding these songs.

2. LORD OF KEDAR

1. "Yahāṃ to keval ek trikoṇ Akār meṃ ūṃcāī vālā sthān hai" (*Dvādaś Jyotirliṅg,* n.d., 36–38).

2. "Koi particular rūp se mān ke nahīṇ calenge."

3. I am grateful to Udi Halperin and David Haberman for drawing me to this way of thinking about "presence."

4. In developing this insight I am deeply indebted to the work of Benjamin Fleming (2007), particularly his dissertation on the relationship of the development of the *jyotirlinga* system to the early history of the forms through which Shiva was worshipped.

5. Kedarnath also appears on other kinds of Shaiva lists of Shiva-oriented worlds and places during the sixth to ninth century CE (Bisschop 2006, 25–35). In the *ayatana* lists, or the lists of places that mark the place, temple, or abode of a deity found in certain versions of the *Skandapurana* manuscripts, Kedarnath is the second Shaiva *ayatana* in the category of "mundane place, accessible to men" (Bisschop 2006, 13). Jan Gonda's (1975, 178–256) discussion of the semantic range of the word *ayatana* offers insight into the difficulties of finding the boundaries between god, form of deity used for worship, and location. On this point, see also Bisschop (2004, 67).

6. The translation is mine. I adopted the suggestion of Peter Bisschop that Nanda might be a river name, and I am grateful to him for his guidance in working with this passage.

7. Several centuries later, the water at Kedara is also connected to the lake of Shiva's potent *retas* (Sanskrit: *retodaka*, semen or mercury or quicksilver) from which Karttikeya was born. On this point, see Bisschop (2006, 181–82) and White (1996, 245–46).

8. *Tevāram* (Campantar 2.114.11). I am indebted to Anne Monius for the reference and translation.

9. Anne Feldhaus (2003, 245) has observed that "the exact list varies."

10. Matthew Clark (2006, 190) has observed that the Kalamukhas began to be associated with specific temples at Baḷḷigāve in Karnataka in 1019. Among these is a temple notably dedicated to the "Dakṣiṇa-Kedāreśvara" (Lord of the Southern Kedāra).

11. It is of note in this regard that there is a *rawal* in the traditionally attested lineage of Kedarnath *rawal-jagatgurus* named, perhaps, for this form of Bhairava: Shri Rawal Bhukundalinga Jagadguru (Hiremath 2006, 62).

12. Some historians of Garhwal are comfortable connecting this story to Garhwal specifically. See Saklani (1998, 40, 80) and Negi (2001, 5–7).

13. For a summary of this point, see M. Joshi (1986).

14. Travis Smith (2007, 89) suggests that the "alliance between Lākulīśa Pāśupatism and Brahmanism runs deeper than most scholars have acknowledged."

15. For a relevant but not identical example of this process, see Chakrabarti (2001, 81).

16. Though it is not strictly possible to absolutely connect the figure of Shankara to the building of a temple at Kedarnath, historian Shivprasad Naithani does deem it possible that a temple at Kedarnath could have been built beginning roughly at this time, probably constructed by the ruling regional dynasty of the time in Garhwal-Kumaon, the Katyuris. On Shankara's presence in the Himalaya, see Pande (1994, 340–50); Bader (2000, 136–82, 252); and Naithānī (2006, 167). I am indebted to Jack Hawley for bringing Jonathan Bader's work to my attention.

17. For how the Garhwali story might fit into the broader history of the Naths, see Mallinson (2009, 421).

18. Colonial records support, albeit in a confusing and sometimes contradictory way, the presence of Virashaivas in the Kedarnath valley in the nineteenth century (Traill 1823, 124; Atkinson 1884, 55–57; Walton [1910] 1989, 173; Oakley 1905, 147).

19. This understanding has been presented by Shoba S. Hiremath (2006, 10–11), author of an English history of Kedarnath sanctioned by the five *jagatgurus*. See also on this point, see Reddy (2014, 118).

20. David White (1996, 121) points out that the title *rawal* may have begun as a Pashupata clan name in the eighth century and been absorbed into the social lexicon of the Naths. He derives it from the Sanskrit term *rāja-kula* (lineage of the king). See also Ḍabarāl (n.d., 414–17).

21. I am grateful to Gil Ben-Herut for his guidance in understanding this early history.

22. Fleming notes that there are two very different accounts of Kedarnath as a *jyotirlinga* in the *Shiva Purana*, one in the *koṭirudrasaṃhita* and one in the *jñānasaṃhita*, that seem to have different understandings of the form into which Shiva is descending. The translation used here is from the influential translation series *Ancient Indian Tradition and Mythology*. The Sanskrit versions of the *Shiva Purana* most commonly available in the Kedarnath valley (they are usually ordered from Haridwar or Rishikesh) are those published by the publisher who took over the Venkateshwar Press, Khemaraja Shri Krishnadas. See Fleming (2007, 13–14, 81–84); Shastri (1970, *Koṭirudrasaṃhita* 19.1–26); *Śrī Śivamahāpurāṇa* (2004).

23. "Saṃsāratāraṇaṃ cā'nya pāpajālanikṛtanam || Kedāramudakaṃ pītvā punarjanma na vidyate| na yoniṣu niyuñjyate sa gacchecchāśvataṃ padam||." Lakshmidhara states that the verse is found in the Devi Purana.

24. Unless otherwise noted, the term *Kedarakhanda* refers to the Garhwal-specific version of this text that is not part of Mahapuranic *Skandapurana* texts. My analysis is based on the five published editions of the *Kedarakhanda* in my possession (Dwivedī 2001; Kṛṣṇākumār 1993; Nautiyal 1994; *Śrīskandamahāpurāṇānāntargatakedārakhaṇḍaḥ* 1906; Vyāsa 2007). For a related published text that may prove helpful in understanding the redaction process of the *Kedarakhanda*, see *Bṛhad Śrī Badrī Nārāyaṇa māhātmya aura cārodhāma Devaprayāga, Pañcakedāra māhātmya, tathā Gaṅgottarī* (1913).

25. *Kedarakhanda* 40.1–29, ed. Nautiyal (1994, 141–43). The translation is mine.

26. On the history of the *yaksha* cult, see Gail Sutherland (1991). On Garhwal-specific *gandharva* traditions, see Dhasmana (1995) on the reation of the Garhwali deity Jakh to the *yaksha*.

27. This passage is found in *Kedarakhanda* 41.1–55, ed. Nautiyal (1994, 144–47). The translated sentences are from *Kedarakhanda* 41.26–29. I am grateful to Nirmila Kulkarni of the Center for Advanced Sanskrit Studies at the University of Pune for introducing me to this concept and her guidance in interpreting this passage.

28. Vaitaraṇī is the name for a legendary river, a form of the Ganga, that flows through a hell into the world of the ancestors. In this passage it is unclear whether this appellation denotes a specific place or whether this is merely an indirect way of acknowledging that the Mandakini, having joined the Alaknanda at Rudraprayag, joins and becomes the Ganga at Devprayag. See Bakker et al. (2014, 89).

29. There are, to my knowledge, two commercially published editions of this text. One was produced by the well-known Mumbai-based publisher Khemaraja Shri Krishnadas, and the other is by the regionally well-known (if you are from Uttarakhand) publisher Vishalmani Sharma, based in the village of Narayan Koti in the Kedarnath valley. The Khemaraja edition begins with several chapters of material that describe ritual preparation preliminary to the physical journey, whereas the Vishalmani Sharma version does not. See Padumā and Hajārībāg (1907); Viśālmaṇi Śarmā Upādhyāy (1952). The chapter colophons of the Khemaraja edition state that this text comes from the *Kedārakalpa* section of the *Rudrayāmala Tantra* (one of the tantric traditions connected to the *mantramarga*). The attribution of the *Kedārakalpa* to the *Rudrayamala* is itself suspect. The *Rudrayamala Tantra*

is the sort of text to which many passages are attributed. See Biernacki (2007, 49), who calls it "that most elusive and ubiquitous of texts"; Muller-Ortega (1989, 6, 42).

30. "Pibeta vṛsabho yathā." See *Kedarakalpa* 6.17, ed. Viśālmaṇi Śarmā Upādhyāy (1952, 27).

31. The drinking of water is also one of the main elements of a passage about Kedarnath from a *dharmashatra* digest that describes what to do when one goes to various pilgrimage places. See Vīramitrodayaḥ, who references the *Devi Purana* (1917, 490–92). On the ubiquity of animal-related vows, see Samuel (2008, 162).

3. EARLIER TIMES

1. On the spatialization of capital, see N. Smith (2011); Prudham and Heynen (2011); N. Smith (2008). I owe this understanding to Sandeep Banerjee, who first suggested I explore Neil Smith's notion of "uneven development."

2. On this point, see also Sax (2011, 167–81).

3. Sudarshan Shah was a king of the Pal (Parmar) dynasty whose rule in Garhwal was consolidated by Ajay Pal. At a certain point in the history of this dynasty the kings assumed the title of Shah (Rawat 2002, 79, 36).

4. Regarding deity ownership of land in the neighboring context of Himachal Pradesh, see Berti (2009).

5. See, among others, Guha (2000, 62–137); Mawdsley (1999); Rangan (2000); Mukherjee (2012).

6. The quoted version is from Husain (1965, 523). I am grateful to James Lochtefeld for the reference.

7. The metallurgical detail here is suggestive of what David White (1996) has observed about the alchemical significance of Kedarnath for Nath-Siddhas.

8. Nivedita's (1928) observations about Agastmuni (27–29), Nāla (33), and Ukhimath (48–49) are notable as well.

9. For an extended critique of this idea, see Ives (2004) and Ives and Messerli (1989), as well as a more extended discussion of this point in chapter 6.

10. It is of course important to recognize that such exponential growth is not unique to this particular Himalayan Indian region. Mahesh Sharma (2009, 113) observes, by comparison, that the number of visits to the Shaiva pilgrimage place Manimahesha in Himachal Pradesh grew from thirty thousand in 1998 to one hundred thousand in 2006.

11. I am grateful to Andrea Pinkney for her suggestion to think of these developments as a form of branding.

4. THE SEASON

1. Unlike most offerings to deities, which are received back as *prasad* and consumed if edible, this rice is never eaten by anyone and is disposed of outside the boundaries of Kedarnath. (Some link this festival to the story in which the gods, when churning the ocean, inadvertently produced a poison that could wipe out all creation but Shiva drank it to save the world and closed his throat to prevent it from descending into his body.)

2. The importance of Kedarnath and more broadly the Himalaya in the Bengali imaginaire receives more substantive discussion in chapter 6.

3. Sanskritists who began the Lanman reader with the story of Nala and Damayanti will be fascinated to note that the village of Nala connects itself to this story.

4. For examples of Garhwali deity processions, see Sax (2000, 102–9; 2006, 119). See also forthcoming work by Karin Polit. On the relationship of the metal masks to the *doli*, and thereby to the deity, see Daniela Berti's (2004) discussion of a similar phenomenon in Himachal Pradesh. On the complex relationships among deity procession, territory, political authority, and ritual in the western Himalaya, see Peter Sutherland (2006, 1998). I am grateful to Udi Halperin for his suggestions about this section about Nala Devi.

5. The Kedarnath *deora* procession also did not involve a *doli* prior to the 1950s.

6. Weapons and other insignia of deities are often in Garhwal regarded as powerfully charged objects. In the context of the *Pandav Nrtya* performance tradition, each character dances with a specially designated weapon that is regarded as full of *shakti*. Sometimes possession begins when someone grasps the weapon or object. It therefore stands to reason that such objects could also be "recharged." See Sax (2002, 84, 97–100).

7. The *tirth purohit* who did these actions was also from Nala village, which may explain his role here.

8. "Buḍha Kedār." There is also a Buḍha Madmaheśvar located on a hill above the valley in which the current Madhyamaheshwar temple is found. That place, however, actually features a very small and simple shrine with a *linga* inside. Additionally, there are other "Kedars" not part of the five-Kedar series: Basukedār (located south of Lamgaundi, arguably the beginning point of the Kedarnath valley pilgrim trail of centuries past) and another Buḍha Kedār, which I have not seen, reputedly on the path that departs from Triyugi Narayan and crosses the mountains that divide the Kedar valley from the Uttarkashi district.

9. In 2007 approximately fifteen to twenty different Garhwali *devtas*, either singly or in groups, visited Kedarnath. Many of these *devtas* visited from the Uttarkashi district of Garhwal, such as the group of *devtas* (Rudreshvar Mahadev, Narsingh, Someshvar, Madri-Devi, Draupadi-Devi) who arrived in Kedarnath on July 29, 2007. I was told that this particular Rudreshvar Mahadev was in fact the *kul-devta* (lineage deity) of sixty-five villages and was taken on *deora* by different groups from among those villages on different years, sometimes consecutively. Jeetu Bagdwal came from the Uttarkashi side for the first time ever in 2008 as part of a group of *devtas* of whom the main deity was Huneshvar Mahadev.

10. *Ghī māliś, ghī lep, ghī lepan, ghṛt lepan*, in order of increasingly Sanskritized Hindi, are all words that refer to the action in question. Of these, *ghī māliś*, or *ghee malish*, is the most common expression. I therefore use *ghee malish*.

11. North Indian *yatris*, without correlation to a particular region, would occasionally tell me that touching and massaging of the *linga* were practices that occurred at certain temples in their home region. Paul Muller-Ortega (personal communication, June 2009) has witnessed a select group of devotees massaging the *linga* at Mallikarjuna, a *jyotirlinga* located in Andhra Pradesh. However, in many other North Indian settings, and in most South Indian settings of which I am aware, such an act would not be possible for the average devotee.

12. "Pāpo'haṃ pāpakarmāhaṃ pāpātmā pāpasambhavaḥ | Trahī māṃ Pārvatīnātha sarvapāpaharo bhava||."

13. This mantra is different from the mantras for the *ghee* bath used during an *abhishek*, and the *ghee malish* is always done whether or not there is an *abhishek*. I am grateful to

members of the Kedarnath Tirth Purohit Association and the Samiti for their explanations of the *ghee malish*.

14. It is also possible to translate this phrase as "blown-out *darshan*," referencing both the idea that the *darshan* bestows a liberation (often translated as *moksha* or often in Buddhist contexts *nirvana*) that can be understood as the extinguishing of *karma* but perhaps also referencing the contrast with the evening *darshan*, which involves fire and light, and the morning *darshan*, which does not.

15. For a fuller discussion of the *ghee* massage in Kedarnath, see Whitmore (2018).

16. In important ways, a video like this should be understood as a twenty-first-century *mahatmya*, a text that aims that aims to present the greatness (Sanskrit: *mahatmhya*) of a particular place, region, and/or deity. On this point, see Pinkney (2013a) and Whitmore (2016).

17. My thinking about the experience of weather in Kedarnath owes a great deal to my many fruitful conversations with Tori Jennings. On the experience of weather, see Jennings (2016).

18. Excerpted and summarized from conversation in Kedarnath, June 6, 2007.

19. In 2007–8 I did not observe any *payari geet* sung during the Kedarnath procession. This to me was a significant difference. Two valleys to the west there was an active tradition of welcoming a traveling deity with song, whereas in the Kedarnath valley this tradition was not in evidence and I heard conflicting reports about whether such songs had ever been sung to Kedarnath. This made sense because Kedarnath-Shiva was important beyond the region and did not function as the guardian deity or owner of a specific area in the manner of the other Shivas of the Panch Kedar. The entire region was his, but not in a way that emphasized his everyday presence and activity.

20. I do not know if any of these hypotheticals have happened since 2007. I have been reluctant to inquire because it is a sensitive matter. D. R. Purohit's vast knowledge of Garhwali literature and practice was immensely helpful to me in contextualizing my own data on these points.

5. WHEN THE FLOODS CAME

1. I am grateful to the anonymous reader provided by the University of California Press for guidance in thinking about the "*longue durée*" of the 2013 floods.

2. D. P. Dobhal, Dehra Dun, personal communication, May 2014.

3. Kedarnath valley resident who survived the floods in Kedarnath, personal communication, May 2014.

4. Shankar Vishwanath, "'Trasadī—Sarkār—Nyūz Cainal—Gaṛhvālī'" [Tragedy—government—news channel—Garhwali], Facebook Wall, July 1, 2013, https://www.facebook.com/shankar.vishwanath.7?lst=2615540%3A100000522888362%3A1530226362.

5. Shashank Srivastava, personal communication, February 6, 2015.

6. NATURE'S TANDAVA DANCE

1. I am grateful to Jonathan Greene for his assistance in thinking through this formulation.

2. The complexity of responses to the floods follows to some degree the model set out by Judith Schlehe (2010) in her work on recent disasters in Indonesia.

3. The well-acknowledged problem of Himalayan mountain hazards is to be distinguished from supposed threats facing the Himalaya presented by the once-regnant "Theory of Himalayan Environmental Degradation." *Himalayan Perceptions* (Ives 2004) was the follow-up project to Ives and Messerli's earlier work *The Himalayan Dilemma* (1989), which questioned the then commonly held idea that the Himalaya was a fragile landscape in environmental crisis because of the cumulative impact of overpopulation and the resultant higher levels of subsistence farming that had resulted in large-scale deforestation and had thereby increased the impact of flooding. This position, termed the Theory of Himalayan Environmental Degradation (THED), carried with it a pervasive discourse about fragility and vulnerability that still infuses popular sentiment since 2013 in Uttarakhand and across Himalayan regions in South Asia, particularly in Nepal. Ives and Messerli's rebuttal of this position has since been ratified by scholarly consensus (Mathur 2015, 102; Ives 2006; Ives and Messerli 1989; Ives 2004; Guthman 2002). More recent work often resists a straightforward fragility narrative or "vulnerability-resilience" binary when thinking about human-nature relationships in the Himalaya. While still seriously attending to the particular forms of hardship that attend residence in the Himalaya, recent scholarship presents a more granular and nuanced focus on issues such as local and regional practices of resource use and management, how those practices have affected floods and sediment flow, and "how Himalayan communities conceptualize their environment" (Wasson et al. 2008; Rangan 1995; Guneratne 2010, 1–3). There has also been a push to "decolonize the Anthropocene" and its sense of anthropogenic environmental crisis and "move beyond a politics of urgency to examine the slow, historical processes of erasure under colonialism and imperialism" (Gergan 2017, 490; see also Bauer and Bhan 2018). On this point, see also recent excellent work by Ritodhi Chakraborti (2018), to whom I am grateful for many insights about this topic. It emphasizes the utility of the idea of "well-being" as a replacement for the problematic vulnerability/resilience binary (2018).

4. On this point, see Gardner (2015).

5. I am grateful to Samantha Kaplan for her help in thinking through the relationships of modern built environments to the natural environments where they are situated.

6. I am grateful to Devaleena Das for this reference.

7. Vijay Ballabh Bhatt, "Mitron Maut Ke Kareeb," Facebook Wall, June 21, 2013, https://www.facebook.com/vijayballabh.bhatt. Where not indicated as English, the language in the original post was in Hindi. The translation is mine.

8. I am grateful to James Lochtefeld for insight about this point.

9. See also a forthcoming book project on this topic by Brian Pennington, tentatively entitled *God's Fifth Abode: Entrepreneurial Hinduism in the Indian Himalayas.*

10. For a complementary perspective on this point see Sax (2011).

11. The influence of European imaginings of the mountainous picturesque that often accompanied the expectations of the tourist gaze in the region is immediately evident in the Kedarnath valley today. The area in and around Tungnath, one of the Panch Kedar that lies on the Chopta road to Badrinath, is known locally as the Switzerland of India.

12. Linkenbach (2006, 165) goes on to note that Uttarakhandi regionalism seems to be developing "anti-Muslim rhetoric." For broader context on this point, see also Cederlöf and Sivaramakrishnan (2006, 31–33); Moore (2003, 187).

13. In 2007 while I was living in Kedarnath several people said to me during informal conversation that in recent years, because of the increased number of *yatri* and amounts of

circulating money, the environment in Kedarnath had become increasingly intense (Hindi: *tez*) and less enjoyable.

14. I am indebted to James Lochtefeld for conversation that aided me in thinking through this situation.

15. M. M. Dhasmana (1995, 44), writing about the worship of the deity Jakh in the Kedarnath valley, confirms this point.

16. On the idea of a "new and holistic relationship between economics and ecology," see Bandyopadhyay and Shiva (1988, 1227).

17. A team of researchers led by Changrae Cho (C. Cho et al. 2016, 797), for example, concluded that "a regional modeling diagnosis attributed 60–90% of rainfall amounts in the June 2013 event to post-1980 climate trends" that could potentially be correlated to "increased loading of green-house gases and aerosols." On this point, see also Agnihotri et al. (2017). I am grateful to Samantha Kaplan for her guidance in understanding the state of scholarly conversation among climate scientists on this point and to Ritodhi Chakraborty for discussion about the Uttarakhand climate situation specifically.

18. As an aside, it should be noted that this report is remarkable in its style and tone. It is characterized by a persistent mingling of what in many circumstances would be separate genres: the scientific languages of hydrology and geology, the social scientific language of environmental anthropology, and the celebratory religious language that I associate with oral and written *mahatmya* accounts. It shifts easily between nuanced accounts of pilgrimage tourism in the region and technical assessments of the impact of hydroelectric dams on local biodiversity. The preface begins with a quote attributed to the *Kedarakhanda* section of the *Skandapurana*, and the second-to-last chapter closes with a quote about the Himalaya from the famous play *Kumarasambhava* by Kalidasa. While I have not focused on it in this book, the religious texture of the biodiversity of the central Himalaya is a subject that warrants consideration in its own regard, beginning with the famous episode in the *Ramayana* (mentioned to me in Kedarnath) when Hanuman carries an entire Himalayan mountain to the southernmost part of the subcontinent so that a wounded Lakshman can be saved by Himalayan plants.

19. See also on this point Drew (2014b, 33); Alley (2000).

7. TOPOGRAPHIES OF REINVENTION

1. I am grateful to Rikhil Bhavnani for clarifying conversations on this point and to Peter Valdina for conversation that helped me to conceptualize this chapter as a whole. It should be noted Emma Mawdsley (2005) has observed, in the context of her work on the Tehri Dam, the complexity of the ways in which the "Hindu Right" has sought involvement with environmental activism. On this point, in the context of postearthquake Gujarat, see Bhattacharjee (2016).

2. On the idea of local deities as entities involved in politics, see Jassal (2016).

3. For other recent examples of this trend toward holistic, multidisciplinary approaches to the study of ecology and religion, see Kent (2010); Veldman, Szasz, and Haluza-DeLay (2012); Frisk (2015); Snodgrass and Tiedje (2008); Drew (2017).

4. NIM staff, Sonprayag, personal communication, January 2017.

REFERENCE

ABP News. "ABP News Enters inside the Temple of Kedarnath." YouTube, June 23, 2013. https://www.youtube.com/watch?v=caoQDSBa3ow&feature=youtube_gdata_player.

Acharya, Diwakar. 2009. "Pāśupatas." In Jacobsen et al. 2009–15, vol. 1.

Agnihotri, Rajesh, A. P. Dimri, H. M. Joshi, N. K. Verma, C. Sharma, J. Singh, and Y. P. Sundriyal. 2017. "Assessing Operative Natural and Anthropogenic Forcing Factors from Long-Term Climate Time Series of Uttarakhand (India) in the Backdrop of Recurring Extreme Rainfall Events over Northwest Himalaya." *Geomorphology* 284 (May): 31–40. https://doi.org/10.1016/j.geomorph.2016.10.024.

Agrawal, Arun, and K. Sivaramakrishnan, eds. 2003. *Regional Modernities: The Cultural Politics of Development in India.* Stanford, CA: Stanford University Press.

Aiyar, Shankar. 2012. *Accidental India: A History of the Nation's Passage through Crisis and Change.* New Delhi: Aleph Books.

Alley, Kelly. 2000. "Separate Domains: Hinduism, Politics, and Environmental Pollution." In *Hinduism and Ecology: The Intersection of Earth, Sky, and Water,* edited by Christopher Key Chapple and Mary Evelyn Tucker, 355–87. Cambridge, MA: Harvard University Press.

Allocco, Amy L. 2009. "Snakes, Goddesses, and Anthills: Modern Challenges and Women's Ritual Responses in Contemporary South India." PhD diss., Emory University.

Alter, Stephen. 2001. *Sacred Waters: A Pilgrimage up the Ganges River to the Source of Hindu Culture.* New York: Harcourt.

Amar Ujala Staff. 2013a. "Kedārnāth meṃ ab ek sāl bād hogī śavoṃ kī talāś." *Amar Ujala,* November 16. www.dehradun.amarujala.com/news/dehradun-local/city-hulchul-dun/kedarnath-disaster-dead-body/.

———. 2013b. "Sab jānanā cāhte hain, kyā huā thā Kedārnāth meṃ?" *Amar Ujala,* October 8. www.amarujala.com/news/states/uttarakhand/every-one-wants-to-know-kedarnath-disaster-story/.

Appadurai, Arjun. 1981. *Worship and Conflict under Colonial Rule: A South Indian Case.* Cambridge: Cambridge University Press.

Aron, Sunita. 2013. "I Am Answerable to My Conscience: Vijay Bahuguna." *Hindustan Times*, June 24. www.hindustantimes.com/india/i-am-answerable-to-my-conscience-vijay-bahuguna/story-aqUqE2NQzxH502cEbDQApM.html.

Atkinson, Edwin T. 1884. "Notes on the History of Religion in the Himálaya of the N.W. Provinces, Part I." *Journal of the Asiatic Society of Bengal* 53, pt. I: 1–103.

———. 1886. *The Himalayan Districts of the North-Western Provinces of India.* Vol. 3. Allahabad: North-Western Provinces and Oudh Government Press.

Bader, Jonathan. 2000. *Conquest of the Four Quarters: Traditional Accounts of the Life of Śaṅkara.* New Delhi: Aditya Prakashan.

Badri-Kedar Temple Committee. "Badri-Kedar Temple Committee Download." n.d. Badarikedar.org. Accessed August 16, 2015. www.badarikedar.org/BKTC/download.

Bagla, Pallava. 2013. "Uttarakhand: The Making of the 'Himalayan Tsunami.'" NDTV.com, July 18. www.ndtv.com/article/india/uttarakhand-the-making-of-the-himalayan-tsunami-393901.

Bahuguna, Sunderlal. 2013. "When Has Rural People's Voice Ever Been Heard?" *Outlook India*, July 8. www.outlookindia.com/article/When-Has-Rural-Peoples-Voice-Ever-Been-Heard/286501.

Bakker, Hans. 1997. *The Vākāṭakas: An Essay in Hindu Iconology.* Groningen: Egbert Forsten.

———. 2014. *The World of the Skandapurāṇa: Northern India in the Sixth and Seventh Centuries.* Leiden: Brill.

Bakker, Hans, Peter C. Bisschop, Yuko Yukochi, Nina Mirnig, and Judit Törzsök, eds. 2014. *The Skandapurāṇa.* Vol. 2b. Leiden: Brill.

Bandyopadhyay, Jayanta, and Vandana Shiva. 1988. "Political Economy of Ecology Movements." *Economic and Political Weekly* 23 (24): 1223–32.

Banerjee, Sandeep. 2014. "'Not Altogether Unpicturesque': Samuel Bourne and the Landscaping of the Victorian Himalaya." *Victorian Literature and Culture* 42 (3): 351–68. https://doi.org/10.1017/S1060150314000035.

Banerjee, Sandeep, and Subho Basu. 2015. "Secularizing the Sacred, Imagining the Nation-Space: The Himalaya in Bengali Travelogues, 1856–1901." *Modern Asian Studies* 49 (3): 609–49. http://dx.doi.org.ezproxy.library.wisc.edu/10.1017/S0026749X13000589.

Barnard, Patrick L., Lewis A. Owen, Milap C. Sharma, and Robert C. Finkel. 2001. "Natural and Human-Induced Landsliding in the Garhwal Himalaya of Northern India." *Geomorphology* 40:21–35.

Basu, Soma. 2013. "Survivors Walk Up to 90 Km to Reach Safety." *Down to Earth* (Nature Conservancy), June 20. www.downtoearth.org.in/content/survivors-walk-90-km-reach-safety.

Bauer, Andrew M., and Mona Bhan. 2018. *Climate without Nature: A Critical Anthropology of the Anthropocene.* Cambridge: Cambridge University Press.

Bayly, C. A. 1988. *Indian Society and the Making of the British Empire.* New Cambridge History of India 2/1. Cambridge: Cambridge University Press.

BBC News India Staff. 2013a. "India Flood Death Toll 'Passes 500.'" BBC News, June 21. www.bbc.co.uk/news/world-asia-india-23007121.

———. 2013b. "Indian Media: Poor Flood Management." BBC News, June 19. www.bbc.co.uk/news/world-asia-india-22963970.

Ben-Herut, Gil. 2015. "Figuring the South-Indian Śivabhakti Movement: The Broad Narrative Gaze of Early Kannada Hagiographic Literature." *Journal of Hindu Studies* 8 (3): 274–95. https://doi.org/10.1093/jhs/hiv025.

———. 2016. "Things Standing Shall Move: Temple Worship in Early Kannada Śivabhakti Hagiographies." *International Journal of Hindu Studies* 20 (2): 129–58. https://doi.org/10.1007/s11407-016-9188-3.

Berreman, Gerald Duane. 1972. *Hindus of the Himalayas: Ethnography and Change*. 2nd ed., rev. and enl. Berkeley: University of California Press.

Berry, Evan. 2014. "Religion and Sustainability in Global Civil Society." *Worldviews: Global Religions, Culture, and Ecology* 18 (3): 269–88.

Berti, Daniela. 2004. "Of Metal and Clothes: The Location of Distinctive Features of Indian Iconography." In *Images in Asian Religions: Texts and Contexts*, edited by Phyllis Granoff and Koichi Shinohara, 85–114. Vancouver: University of British Columbia Press.

———. 2009. "Divine Jurisdictions and Forms of Government in Himacal Pradesh." In *Territory, Soil, and Society in South Asia*, edited by Daniela Berti and Gilles Tarabout, 311–40. New Delhi: Manohar.

———. 2015. "Gods' Rights vs Hydroelectric Projects: Environmental Conflicts and the Judicialization of Nature in India." In *The Human Person and Nature in Classical and Modern India*, edited by R. Torella and G. Milanetti, 111–29. Alla Rivista Degli Sudi Orientali, suppl. 2. Pisa: Fabrizio Serra Editore.

———. 2016. "Plaintiff Deities: Ritual Honours as Fundamental Rights in India." In *Filing Religion: State, Hinduism, and Courts of Law*, edited by Gilles Tarabout and Raphaël Voix, 71–100. New Delhi: Oxford University Press.

Bhandari, Preety M., and Charlotte Benson. 2013. "Live Online Chat—Facing Nature's Wrath: Dealing with Climate Change and Its Effects." Asian Development Bank event, November 28. www.adb.org/news/events/live-online-chat-facing-natures-wrath-dealing-climate-change-and-its-effects.

Bhardwaj, Surinder Mohan. 1973. *Hindu Places of Pilgrimage in India*. Delhi: Munshiram Manoharlal.

Bhardwaj, Surinder Mohan, and J. G. Lochtefeld. 2004. "Tirtha." In *The Hindu World*, edited by Sushil Mittal and Gene R. Thursby, 478–501. New York: Routledge.

Bhaskar News Staff. 2013. "Uttarakhand Floods: Lack of Wood, Controversy Theories and Rituals Hinder Mass Cremation in Kedarnath." *Daily Bhaskar*, June 27. http://daily.bhaskar.com/article/CHD-uttarakhand-floods-lack-of-wood-controversy-theories-and-rituals-hinder-mass-cre-4303606-NOR.html.

Bhatt, Abhinav. 2013. "300 People Still Stranded in Uttarakhand, Awaiting Rescue." NDTV.com, July 4. www.ndtv.com/article/india/300-people-still-stranded-in-uttarakhand-awaiting-rescue-387736.

Bhatt, Abhinav, and Press Trust of India. 2013. "Uttarakhand: Over 1000 Still Stranded, Rescue Operations Continue." NDTV.com, June 29. www.ndtv.com/article/cheat-sheet/uttarakhand-over-1000-still-stranded-rescue-operations-continue-385531.

Bhatt, Brijesh. 2013. "Aur itihās ban gaī Kedārnāth yātrā." *Dainik Jagran*, November 6. www.jagran.com/news/national-kedarnath-yatra-became-history-after-tragedy-10843475.html.

Bhatt, Jagdish. 2013. "Modi's Gesture to Flood-Hit Uttarakhand." *Hill Post*, July 13. http://hillpost.in/2013/07/modis-gesture-to-flood-hit-uttarakhand/93766/.

———. 2014. "Will the Char Dhaam Yatra Be Written Off This Season?" *Hill Post*, February 18. http://hillpost.in/2014/02/will-the-char-dhaam-yatra-be-written-off-this-season/98044/.

Bhattacharjee, Malini. 2016. "Sevā, Hindutva, and the Politics of Post-earthquake Relief and Reconstruction in Rural Kutch." *Asian Ethnology* 75 (1): 75–104.

Bhattacharyya, Debjani. 2017. "Being, River: The Law, the Person and the Unthinkable." *H-Law* (blog), April 26, 2017. https://networks.h-net.org/node/16794/blog/world-legal-history-blog/177310/being-river-law-person-and-unthinkable.

Biernacki, Loriliai. 2007. *Renowned Goddess of Desire: Women, Sex, and Speech in Tantra.* Oxford: Oxford University Press.

Bilham, R. 2004. "Earthquakes in India and the Himalaya: Tectonics, Geodesy and History." *Annals of Geophysics* 47 (2–3). https://doi.org/10.4401/ag-3338.

Bisschop, Peter Christiaan. 2004. "Śiva's Āyatanas in the Various Recentions of Skandapurāṇa 167." In *Origin and Growth of the Purāṇic Text Corpus with Special Reference to the Skandapurāṇa,* edited by Hans T. Bakker, 65–78. Delhi: Motilal Banarsidass.

———. 2006. *Early Śaivism and the Skandāpurāṇa: Sects and Centres.* Groningen Oriental Studies. Groningen: E. Forsten.

———. 2009. "Śiva." In Jacobsen et al. 2009–15, vol. 1.

———. 2010. "Śaivism in the Gupta-Vākāṭaka Age." *Journal of the Royal Asiatic Society,* 3rd ser., 20 (4): 477–88.

Bṛhad Śrī Badrī Nārāyaṇa māhātmya aura cārodhāma Devaprayāga, Pañcakedāra māhātmya, tathā Gaṅgottarī. 1913. Jilā-Gaḍhavāla: Maheśānandaśarmā.

Brosius, Christiane. 2010. *India's Middle Class: New Forms of Urban Leisure, Consumption and Prosperity.* New Delhi: Routledge.

Casey, Edward S. 1996. "How to Get from Space to Place in a Fairly Short Stretch of Time: Phenomenological Prolegomena." In *Senses of Place,* edited by Steven Feld and Keith H. Basso, 13–52. Santa Fe, NM: School of American Research Press.

Cederlöf, Gunnel, and K. Sivaramakrishnan, eds. 2006. *Ecological Nationalisms: Nature, Livelihoods, and Identities in South Asia.* Seattle: University of Washington Press.

Chakrabarti, Kunal. 2001. *Religious Process: The Purāṇas and the Making of a Regional Tradition.* New Delhi: Oxford University Press.

Chakraborty, Ritodhi. 2018. "The Invisible (Mountain) Man: Migrant Youth and Relational Vulnerability in the Indian Himalayas." PhD diss., University of Wisconsin–Madison.

Chanchani, Nachiket. 2012. "Fordings and Frontiers: Architecture and Identity in the Central Himalayas (c. 7th–12th Centuries CE)." PhD diss., University of Pennsylvania.

———. 2014. "The Jageshwar Valley, Where Death Is Conquered." *Archives of Asian Art* 63 (2): 133–54. https://doi.org/10.1353/aaa.2014.0000.

———. 2015. "Pandukeshwar, Architectural Knowledge, and an Idea of India." *Ars Orientalis* 45. http://dx.doi.org/10.3998/ars.13441566.0045.002.

Chandramohan, C. K. 2013a. "BJP Criticises Move to Resume Prayers at Kedarnath." *Hindu,* September 4. www.thehindu.com/news/national/other-states/bjp-criticises-move-to-resume-prayers-at-kedarnath/article5090574.ece.

———. 2013b. "Now, Tales of Woe Pour In." *Hindu,* June 22. www.thehindu.com/news/national/other-states/now-tales-of-woe-pour-in/article4838045.ece.

Chapple, Christopher Key. 1998. "Towards an Indigenous Indian Environmentalism." In Nelson 1998, 13–37.

Chau, Adam Yuet. 2008. "The Sensorial Production of the Social." *Ethnos* 73 (4): 485–504. https://doi.org/10.1080/00141840802563931.

Chaujar, Ravinder Kumar. 2009. "Climate Change and Its Impact on the Himalayan Glaciers: A Case Study on the Chorabari Glacier, Garhwal Himalaya, India." *Current Science* 96:703–8.

Chidester, David, and Edward Tabor Linenthal, eds. 1995. *American Sacred Space.* Bloomington: Indiana University Press.

Cho, Changrae, Rong Li, S.-Y. Wang, Jin-Ho Yoon, and Robert R. Gillies. 2016. "Anthropogenic Footprint of Climate Change in the June 2013 Northern India Flood." *Climate Dynamics* 46 (3–4): 797–805. https://doi.org/10.1007/s00382-015-2613-2.

Cho, Francisca, and Richard King Squier. 2013. "Religion as a Complex and Dynamic System." *Journal of the American Academy of Religion* 81 (2): 357–98.

Chopra, Ravi. 2013. "The Untold Story from Uttarakhand." *Hindu,* June 26, 2013. www.thehindu.com/opinion/lead/the-untold-story-from-uttarakhand/article4847166.ece.

———. 2014. *Uttarakhand: Development and Ecological Sustainability.* New Delhi: Oxfam India. https://www.oxfamindia.org/sites/default/files/WP8UttarakhandDevpEcoSustainabiit3.pdf.

Chopra, Ravi, B. P. Das, Hemant Dhyani, Ajay Verma, H. S. Venkatesh, H. B. Vasistha, D. P Dobhal, et al. 2014. *Assessment of Environmental Degradation and Impact of Hydroelectric Projects during the June 2013 Disaster in Uttarakhand.* Ministry of Environment and Forests, Government of India. www.indiaenvironmentportal.org.in/files/file/environmental%20degradation%20&%20hydroelectric%20projects.pdf.

Clark, Matthew. 2006. *The Daśanāmī-Saṃnyāsīs: The Integration of Ascetic Lineages into an Order.* Leiden: Brill.

Coccossis, Harry, and Alexandra Mexa, eds. 2004. *The Challenge of Tourism Carrying Capacity Assessment: Theory and Assessment.* Aldershot: Ashgate.

Coleman, Simon. 2002. "Do You Believe in Pilgrimage? Communitas, Contestation and Beyond." *Anthropological Theory* 2 (3): 355–68.

Coleman, Simon, and John Eade. 2004. *Reframing Pilgrimage: Cultures in Motion.* London: Routledge.

Corbridge, Stuart. 2004. "Waiting in Line, or the Moral and Material Geographies of Queue-Jumping." In *Geographies and Moralities: International Perspectives on Development, Justice, and Place,* edited by Roger Lee and David M. Smith, 183–98. Malden, MA: Blackwell. https://doi.org/10.1002/9780470753057.ch12.

Crown Representative's Records: Political Department Records. 1936. "Tehri-Garhwal Darbar's Claim for the Kedarnath and Badrinath Shrines." IOR/R/1/1/2951, India Office Library, London.

Ḍabarāl, Śivaprasād. n.d. *Śrī Uttarākhaṇḍ yātrā darśan.* Nārayaṇ Koṭi, Gaṛhvāl. Viśālmaṇi Śarma Upadhyāya.

Daily Bhaskar Staff. 2013. "Uttarakhand Floods: 1,04,095 People Rescued, 'Operation Rahat' Biggest in the World." *Daily Bhaskar,* June 28. http://daily.bhaskar.com/news/CHD-uttarakhand-floods-104095-people-rescued-operation-rahat-biggest-in-the-world-4304685-NOR.html.

Damon, Frederick H., and Mark S. Mosko, eds. 2005. *On the Order of Chaos: Social Anthropology and the Science of Chaos.* New York: Berghahn Books.

Daniel, E. Valentine. 1984. *Fluid Signs: Being a Person the Tamil Way*. Berkeley: University of California Press.

Dave, Kapil. 2013. "Modi Requests for Special Trains to Gujarat from Uttarakhand." *Times of India*, June 20. http://timesofindia.indiatimes.com/india/Modi-requests-for-special-trains-to-Gujarat-from-Uttarakhand/articleshow/20679530.cms.

Davis, Richard H. 1991. *Ritual in an Oscillating Universe: Worshiping Śiva in Medieval India*. Princeton, NJ: Princeton University Press.

———. 1995. "The Origin of Linga Worship." In *Religions of India in Practice*, edited by Donald S. Lopez, 637–48. Princeton, NJ: Princeton University Press.

Descola, Philippe. 2013. *Beyond Nature and Culture*. Translated by Janet Lloyd. Chicago: University of Chicago Press.

Desjarlais, Robert. 2003. *Sensory Biographies: Lives and Deaths among Nepal's Yolmo Buddhists*. Berkeley: University of California Press.

Desjarlais, Robert, and C. Jason Throop. 2011. "Phenomenological Approaches in Anthropology." *Annual Review of Anthropology*, no. 40:87–102.

Dhasmana, M. M. 1995. "Perception of Bhutas in Kedarkhand." In *Primal Elements: The Oral Tradition*, edited by Baidyanath Saraswati, 1:39–54. New Delhi: Indira Gandhi National Centre for the Arts.

Disaster Mitigation and Management Centre, Government of Uttarakhand, India. 2015. "NGO Works." Accessed April 22. http://dmmc.uk.gov.in/pages/view/75-ngo-works.

Dobhal, D. P, A. K. Gupta, M. Mehta, and D. D. Khandelwal. 2013. "Kedarnath Disaster: Facts and Plausible Causes." *Current Science* 105 (2): 171–74.

Down to Earth Staff. 2013. "Flood Fury: Writing Was on the Wall." *Down to Earth*, June 20, 2013. www.downtoearth.org.in/content/flood-fury-writing-was-wall.

Drew, Georgina. 2014a. "Developing the Himalaya: Development as If Livelihoods Mattered." *Himalaya, the Journal of the Association for Nepal and Himalayan Studies* 34 (2), article 7.

———. 2014b. "A Retreating Goddess? Conflicting Perceptions of Ecological Change Near the Gangotri-Gaumukh Glacier." In *How The World's Religions Are Responding to Climate Change*, edited by Robin Globus Veldman, Andrew Szasz, and Randolph Haluza-DeLay. London: Routledge.

———. 2017. *River Dialogues: Hindu Faith and the Political Ecology of Dams on the Sacred Ganga*. Tucson: University of Arizona Press.

Drew, Georgina, and Ashok Gurung. 2014. "Guest Editors' Introduction: Everyday Religion, Sustainable Environments, and New Directions in Himalayan Studies." *Journal for the Study of Religion, Nature and Culture* 8 (4): 389–404.

Dube, Ishita Banerjee. 2001. *Divine Affairs: Religion, Pilgrimage and the State in Colonial and Postcolonial India*. Delhi: Manohar.

Dundoo, Sangeetha Devi. 2013. "And the Mountains Echoed." *Hindu*, September 2. www.thehindu.com/todays-paper/tp-national/and-the-mountains-echoed/article5083833.ece.

Dvādaś Jyotirliṅg: Bhagavān Śaṃkar ke 12 jyotirliṅgoṃ kī kathā. n.d. Delhi: Lāl Cand and Sons.

Dwivedī, Vācaspati, ed. 2001. *Kedārakhaṇḍa of Maharṣi Vyāsa with the Hindi Commentary*. Translated by Vācaspati Dwivedī. Vol. 1. Gaṅgānāthajhā Granthamālā 17. Varanasi: Sampurnanand Sanskrit University.

Eade, John, and Michael J. Sallnow. 1991. *Contesting the Sacred: The Anthropology of Christian Pilgrimage*. London: Routledge.

Eck, Diana. 1999. *Banaras: City of Light*. New York: Columbia University Press.

——. 2012. *India: A Sacred Geography*. New York: Harmony Books.

Edensor, Tim. 2007. "Sensing the Ruin." *Senses and Society* 2 (2): 217–32. https://doi. org/10.2752/174589307X203100.

——. ed. 2010. *Geographies of Rhythm: Nature, Place, Mobilities and Bodies*. Farnham: Ashgate.

Elmore, Mark. 2016. *Becoming Religious in a Secular Age*. Oakland: University of California Press.

Erndl, Kathleen. 1993. *Victory to the Mother : The Hindu Goddess of Northwest India in Myth, Ritual, and Symbol*. New York: Oxford University Press.

Feener, R. Michael. 2016. "Religion and Reconstruction in the Wake of Disaster." *Asian Ethnology* 75 (1): 191–202.

Feener, R. Michael, and Patrick T. Daly, eds. 2016. *Rebuilding Asia Following Natural Disasters: Approaches to Reconstruction in the Asia-Pacific Region*. Cambridge: Cambridge University Press.

Feld, Steven, and Keith H. Basso, eds. 1996. *Senses of Place*. School of American Research Advanced Seminar Series. Santa Fe, NM: School of American Research Press.

Feldhaus, Anne. 1995. *Water and Womanhood: Religious Meanings of Rivers in Maharashtra*. New York: Oxford University Press.

——. 2003. *Connected Places: Region, Pilgrimage, and Geographical Imagination in India*. Religion/Culture/Critique. New York: Palgrave Macmillan.

Ferguson, James, and Akhil Gupta, eds. 1997. *Culture, Power, Place: Explorations in Critical Anthropology*. Durham, NC: Duke University Press.

Fiol, Stefan. 2010. "Dual Framing: Locating Authenticities in the Music Videos of Himalayan Possession Rituals." *Ethnomusicology* 54 (1): 28–53.

——. 2012. "Articulating Regionalism through Popular Music: The Case of Nauchami Narayana in the Uttarakhand Himalayas." *Journal of Asian Studies* 71 (2): 447–73. https://doi.org/10.1017/S0021911812000101.

Fleming, Benjamin J. 2007. "The Cult of the Jyotirliṅgas and the History of Śaivite Worship." PhD diss., McMaster University.

——. 2009a. "The Form and Formlessness of Śiva: The Linga in Indian Art, Mythology, and Pilgrimage." *Religion Compass* 3 (3): 440–58. https://doi.org/10.1111/j.1749-8171.2009.00141.x.

——. 2009b. "Mapping Sacred Geography in Medieval India: The Case of the Twelve *Jyotirliṅgas*." *International Journal of Hindu Studies* 13 (1): 51–81. https://doi.org/10.1007/s11407-009-9069-0.

Flueckiger, Joyce Burkhalter. 2013. *When the World Becomes Female: Guises of a South Indian Goddess*. Bloomington: Indiana University Press.

Fountain, Philip, and Levi McLaughlin. 2016. "Salvage and Salvation. Guest Editors' Introduction." *Asian Ethnology* 75 (1): 1–28.

Frisk, Kristian. 2015. "Citizens of Planet Earth: The Intertwinement of Religion and Environmentalism in a Globalization Perspective." *Journal for the Study of Religion, Nature and Culture (Online)* 9 (1): 68–86. https://doi.org/10.1558/jsrnc.v9i1.16450.

Gadgil, Madhav, and Ramachandra Guha. 1993. *This Fissured Land: An Ecological History of India*. Berkeley: University of California Press.

——. 1995. *Ecology and Equity: The Use and Abuse of Nature in Contemporary India*. London: Routledge.

Gaenszle, Martin, and Jörg Gengnagel. 2006. *Visualizing Space in Banaras: Images, Maps, and the Practice of Representation.* Ethno-Indology, vol. 4. Wiesbaden: Harrassowitz.

Gaillard, J.C., and P. Texier. 2010. "Religions, Natural Hazards, and Disasters: An Introduction." *Religion* 40 (2): 81–84. https://doi.org/10.1016/j.religion.2009.12.001.

Gamburd, Michele Ruth. 2012. *The Golden Wave: Culture and Politics after Sri Lanka's Tsunami Disaster.* Bloomington: Indiana University Press.

Gandhi, Ajay. 2013. "Standing Still and Cutting in Line." *South Asia Multidisciplinary Academic Journal,* published online March 15. https://samaj.revues.org/3519.

Gardner, James S. 2015. "Risk Complexity and Governance in Mountain Environments." In *Risk Governance: The Articulation of Hazard, Politics, and Ecology,* edited by Urbano Fra Paleo, 349–71. Dordrecht: Springer. https://doi.org/10.1007/978-94-017-9328-5_19.

Gergan, Mabel Denzin. 2015. "Animating the Sacred, Sentient and Spiritual in Post-humanist and Material Geographies." *Geography Compass* 9 (5): 262–75. https://doi.org/10.1111/gec3.12210.

———. 2017. "Living with Earthquakes and Angry Deities at the Himalayan Borderlands." *Annals of the American Association of Geographers* 107 (2): 490–98. https://doi.org/10.1080/24694452.2016.1209103.

Gladstone, David L. 2005. *From Pilgrimage to Package Tour: Travel and Tourism in the Third World.* New York: Routledge.

Gold, Ann Grodzins. 1988. *Fruitful Journeys: The Ways of Rajasthani Pilgrims.* Berkeley: University of California Press.

———. 1998. "Sin and Rain: Moral Ecology in Rural North India." In Nelson 1998, 165–96.

Gold, Ann Grodzins, and Bhoju Ram Gujar. 2002. *In the Time of Trees and Sorrows: Nature, Power, and Memory in Rajasthan.* Durham, NC: Duke University Press.

Gonda, Jan. 1963. "The Indian Mantra." *Oriens* 16 (December): 244–97. https://doi.org/10.2307/1580265.

———. 1975. "Āyatana." In *Selected Studies,* vol. 2, *Sanskrit Word Studies,* 178–256. Leiden: Brill.

Gordon, S. 1985. "Kinship and Pargana in Eighteenth Century Khandesh." *Indian Economic and Social History Review* 22 (December): 371–97.

Goswamy, B.N. 2013. "Pilgrimage." In *Yoga: Art of Transformation,* edited by Debra Diamond and Molly Emma Aitken, 190–95. Washington, DC: Arthur M. Sackler Gallery, Smithsonian Institution.

Govindrajan, Radhika. 2015. "'The Goat That Died for Family': Animal Sacrifice and Interspecies Kinship in India's Central Himalayas." *American Ethnologist* 42 (3): 504–19. https://doi.org/10.1111/amet.12144.

Grim, John, and Mary Evelyn Tucker. 2014. *Ecology and Religion.* Washington, DC: Island Press.

Guha, Ramachandra. 2000. *The Unquiet Woods: Ecological Change and Peasant Resistance in the Himalaya.* Expanded ed. Berkeley: University of California Press.

Guneratne, Arjun. 2010. *Culture and the Environment in the Himalaya.* New York: Routledge.

Gupta, Smita, and P. Sunderarajan. 2013. "Centre Ramps Up Rescue Work in Uttarakhand." *Hindu,* June 22. www.thehindu.com/news/national/centre-ramps-up-rescue-work-in-uttarakhand/article4836934.ece.

Gusain, Raju. 2013a. "Ghee Deluge Threatens to Erode Kedarnath Temple." *India Today*, September 21. http://indiatoday.intoday.in/story/kedarnath-asi-says-ghee-deluge-threatens-temple-uttarakhand/1/325837.html.

———. 2013b. "Uttarkashi-Based Institute Recommends Alternative 34 Km Kedarnath Route to Uttarakhand Govt." *India Today*, July 16. http://indiatoday.intoday.in/story/kedarnath-new-route-uttarakhand-governmentflash-floods-uttarkashi-based-institute/1/291524.html.

———. 2014a. "April Snowfall Hampers Kedarnath Road Repair." *India Today*, April 21. http://indiatoday.intoday.in/story/april-snowfall-hampers-kedarnath-road-repair/1/356758.html.

———. 2014b. "Brand New Kedarnath in the Making." *India Today*, January 10. http://indiatoday.intoday.in/story/brand-new-kedarnath-in-the-making/1/335352.html.

Guthman, Julie. 2002. "Representing Crisis: The Theory of Himalayan Environmental Degradation and the Project of Development in Post-Rana Nepal." *Development and Change* 28 (1): 45–69. https://doi.org/10.1111/1467-7660.00034.

Haberman, David L. 2013. *People Trees: Worship of Trees in Northern India*. New York: Oxford University Press.

Haila, Yrjö, and Chuck Dyke. 2006. *How Nature Speaks: The Dynamics of the Human Ecological Condition*. Durham, NC: Duke University Press.

Halperin, Ehud. 2012. "Haḍimbā Becoming Herself: A Himalayan Goddess in Change." PhD diss., Columbia University. http://academiccommons.columbia.edu/item/ac:178438.

———. 2017. "Winds of Change: Religion and Climate in the Western Himalayas." *Journal of the American Academy of Religion* 85 (1): 64–111. https://doi.org/10.1093/jaarel/lfw042.

Handelman, Don, and Galina Lindquist, eds. 2011. *Religion, Politics, and Globalization: Anthropological Approaches*. New York: Bergahn Books.

Handelman, Don, and David Shulman. 1997. *God Inside Out: Śiva's Game of Dice*. New York: Oxford University Press.

———. 2004. *Siva in the Forest of Pines: An Essay on Sorcery and Self-Knowledge*. Oxford: Oxford University Press.

Hegarty, James. 2012. *Religion, Narrative, and Public Imagination in South Asia: Past and Place in the Sanskrit Mahabharata*. London: Routledge.

Hewitt, K. 1983. "The Idea of Calamity in a Technocratic Age." In *Interpretation of Calamities*, edited by K. Hewitt, 3–32. Risks and Hazards Series 1. Boston: Allen and Unwin.

Hill, Christopher V. 2008. *South Asia: An Environmental History*. ABC-CLIO's Nature and Human Societies Series. Santa Barbara, CA: ABC-CLIO.

Hill Post Staff. 2013. "Kedarnath Valley to Be Fumigated, Says Uttarakhand Minister." *Hill Post*, July 3. http://hillpost.in/2013/07/kedarnath-valley-to-be-fumigated-says-uttarakhand-minister/92590/

Hindu Staff. 2013a. "Monsoon's *Rudratandav* in Uttarakhand." *Hindu*, June 19. www.thehindu.com/news/national/other-states/monsoons-rudratandav-in-uttarakhand/article4830274.ece.

———. 2013b. "'They Charged Rs. 4,000 for a Gas Cylinder.'" *Hindu*, July 3. www.thehindu.com/news/national/karnataka/they-charged-rs-4000-for-a-gas-cylinder/rticle4874294.ece.

Hindustan Times Staff. 2013. "Modi's Uttarakhand Visit Sparks Political War of Words." *Hindustan Times*, June 23. www.hindustantimes.com/india/modi-s-uttarakhand-visit-sparks-political-war-of-words/story-o3mX45Jgs11MMD2JK6in9N.html.

Hiremath, Shobha S. 2006. *Kedar vairagya peetha: Parampara & Rawal Jagadguru Shri Bheemashankarlinga Shivacharya.* Ukhimatha (Ushamath): Himavat Kedar Vairagya Simhasana Mahasamsthana.

Hoffman, Susanna M., and Anthony Oliver-Smith. 1999. "Anthropology and the Angry Earth: An Overview." In Oliver-Smith and Hoffman 1999, 1–16.

Howes, David, and Constance Classen. 2013. *Ways of Sensing: Understanding the Senses in Society.* New York: Routledge.

HT Correspondents. 2013. "Uttarakhand: Death Toll Rising; after Rain, Hunger Begins to Kill." *Hindustan Times*, June 22. www.hindustantimes.com/india/uttarakhand-death-toll-rising-after-rain-hunger-begins-to-kill/story-qAvKnGPWFnRe4ikyMSIeoM.html.

Hubbard, Phil, Rob Kitchin, and Gill Valentine. 2004. *Key Thinkers on Space and Place.* London: Sage Publications.

Huber, Toni. 2008. *The Holy Land Reborn: Pilgrimage and the Tibetan Reinvention of Buddhist India.* Chicago: University of Chicago Press.

Husain, Syed Mohammad. 1965. "The United Provinces Shri Badrinath and Shri Kedarnath Temples Act, 1939 (U.P. Act No. XVI of 1939 as Amended by U.P. Act XXI of 1963 and No. VIII of 1964." In *S.M. Husain's the Uttar Pradesh Local Acts (1793–1964) Annotated: Together with Central Acts with Local Application, Bengal Regulations with Rules, Orders, Regulations and Notifications Etc.*, 4th ed., 1:523–44. Lucknow: Eastern Book Co.

Inden, Ronald B. 1990. *Imagining India.* Oxford: Blackwell.

———. 2011. "Paradise on Earth: The Deccan Sultanates." In *Garden and Landscape Practices in Pre-colonial India,* edited by Emma J. Flatt and Daud Ali, 74–97New Delhi: Routledge.

India-China Institute. n.d. "Everyday Religion and Sustainable Environments in the Himalaya." India-China Institute. Accessed June 24, 2016. www.indiachinainstitute.org/initiatives/everyday-religion/.

India Office. 1895. "Appointment of a Chief Priest, or Naib Rawal, at the Temple of Badrinath in British Garhwal." Judicial and Public Annual Files 1901–1999, IOR/L/PJ/6/408, India Office Library, London.

———. 1942. "File 46 The United Provinces Shri Badri Nath Temple (Amendment) Act, 1942 (United Provinces Act No. XV of 1942)." IOR/L/PJ/7/3389, India Office Records and Private Papers, British Library.

India Today Agencies. 2013. "Kedarnath News: Operation Surya Hope Underway." *India Today*, June 20. http://indiatoday.intoday.in/story/kedarnath-news-operation-surya-hope-underway/1/284309.html.

Ingold, Tim. 2000. *The Perception of the Environment: Essays on Livelihood, Dwelling and Skill.* London: Routledge.

Ivakhiv, Adrian J. 2012. "Religious (Re-)Turns in the Wake of Global Nature: Toward a Cosmopolitics." In *Nature, Science, and Religion: Intersections Shaping Society and the Environment,* edited by Catherine M. Tucker, 213–30. Santa Fe, NM: School for Advanced Research Press.

Ives, Jack D. 2004. *Himalayan Perceptions: Environmental Change and the Well-Being of Mountain Peoples.* London: Routledge.

———. 2006. "Comment: Forests and Floods: Drowning in Fiction or Thriving on Facts." *Mountain Research and Development* 26 (2): 187–88. http://dx.doi.org/10.1659/0276-4741(2006)26[187:FAFDIF]2.0.CO;2.

Ives, Jack D, and Bruno Messerli. 1989. *The Himalayan Dilemma: Reconciling Development and Conservation.* London: Routledge.

Jackson, Michael. 2008. "The Shock of the New: On Migrant Imaginaries and Critical Transitions." *Ethnos* 73 (1): 57–72.

———. 2009. *The Palm at the End of the Mind: Relatedness, Religiosity, and the Real.* Durham, NC: Duke University Press.

Jacobsen, Knut A., Helene Basu, Angelika Malinar, and Vasudha Narayanan, eds. 2009–15. *Brill's Encyclopedia of Hinduism.* 6 vols. Leiden: Brill.

Jagran Post Staff. 2014a. "ADB to Grant Loan of Rs 1,200 Crore for Reconstruction of Uttarakhand." *Jagran Post*, February 6. http://post.jagran.com/adb-to-grant-loan-of-rs-1200-crore-for-reconstruction-of-uttarakhand-1391670155.

———. 2014b. "U'khand CM for Quick Release of Rs 300 Crore from Centre for Road Repair." *Jagran Post*, January 29. http://post.jagran.com/ukhand-cm-for-quick-release-of-rs-300-crore-from-centre-for-road-repair-1390997604.

———. 2014c. "Uma Bharti Appointed Poll in-Charge of Uttarakhand BJP Unit." *Jagran Post*, February 3. http://post.jagran.com/uma-bharti-appointed-poll-incharge-of-uttarakhand-bjp-unit-1391440544.

———. 2014d. "Uttarakhand Government Launches Several Development Projects." *Jagran Post*, February 10. http://post.jagran.com/uttarakhand-government-launches-several-development-projects-1392000073.

———. 2014e. "Uttarakhand Govt to Arrange High Tech Safety Features for Kedarnath Pilgrims." *Jagran Post*, March 10. http://post.jagran.com/uttarakhand-govt-to-arrange-high-tech-safety-features-for-kedarnath-pilgrims-1394432589.

———. 2014f. "Uttarakhand Govt to Provide Compensation to Calamity-Hit Shopkeepers." *Jagran Post*, March 1. http://post.jagran.com/uttarakhand-govt-to-provide-compensation-to-calamityhit-shopkeepers-1393647339.

Jain, Madhu. 1995. *The Abode of Mahashiva: Cults and Symbology in Jaunsar-Bawar in the Mid-Himalayas.* New Delhi: Indus.

James, George Alfred. 2013. *Ecology Is Permanent Economy: The Activism and Environmental Philosophy of Sunderlal Bahuguna.* Albany: State University of New York Press.

Jassal, Aftab Singh. 2016. "Divine Politicking: A Rhetorical Approach to Deity Possession in the Himalayas." *Religions* 7 (9): 117. https://doi.org/10.3390/rel7090117.

Jayaraman, Nityanand. 2013. "A 'Development' That Harms Us All." *Hindu*, July 2. www.thehindu.com/opinion/op-ed/a-development-that-harms-us-all/article4873535.ece.

Jeffrey, Craig. 2010. *Timepass: Youth, Class, and the Politics of Waiting in India.* Stanford, CA: Stanford University Press.

Jenkins, Willis. 2011. "Sustainability." In *Grounding Religion: A Field Guide to the Study of Religion and Ecology,* edited by Whitney Bauman, Richard R. Bohannon, and Kevin J. O'Brien, 96–112. London: Routledge.

Jennings, Tori. 2016. "Cornish Weather and the Phenomenology of Light: On Anthropology and 'Seeing.'" In *Anthropology and Climate Change: From Actions to Transformations,* edited by Susan A. Crate and Mark Nuttal, 2nd ed., 241–49. New York: Routledge.

Joshi, Hridayesh. 2014. *Tum cup kyoṃ rahe Kedār: Himālay kī sabse baṛī trāsdī se uṭhe savāl.* Delhi: Alekh Prakashan.

———. 2016. *Rage of the River: The Untold Story of Kedarnath Disaster.* Translated by Vandana R. Singh. Gurgaon (Haryana), India: Penguin Books India.

Joshi, Maheshwar P. 1986. "Religious History of Uttarakhand: Sources and Materials." In *Ecology, Economy, and Religion of Himalayas,* edited by Lalita Prasad Vidyarthi and Makhan Jha, 193–216. Delhi: Orient Publications.

———. 1990. *Uttaranchal (Kumaon-Garhwal) Himalaya: An Essay in Historical Anthropology.* ASH (Association of Studies on Himalayas) Monographic Series, no. 1. Almora, U.P. Himalayas, India: Shree Almora Book Depot.

———. 2007. "Lakulīśa in the Art and Architecture of Uttarakhand Himalaya." *Journal of Indian Society of Oriental Art,* new ser., 27 (Commemorating 100 Years of Indian Society of Oriental Art): 215–30.

———. 2011. "Geocultural Identities and Belonging in the Ethnohistory of Central Himalaya, Uttarakhand, India." In *The Politics of Belonging in the Himalayas: Local Attachments and Boundary Dynamics,* edited by Joanna Pfaff-Czarnecka and Gérard Toffin, 4: 272–90. Governance, Conflict, and Civic Action Series. New Delhi: Sage Publications.

Joshi, Sandeep. 2014a. "Pilgrims Wary of Undertaking Char Dham Yatra." *Hindu,* April 24. www.thehindu.com/todays-paper/tp-national/pilgrims-wary-of-undertaking-chardham-yatra/article5941893.ece.

———. 2014b. "Technology to Bolster Security for Kedarnath Yatra." *Hindu,* April 25. www.thehindu.com/news/national/other-states/technology-to-bolster-security-for-kedarnath-yatra/article5944614.ece.

Kālidāsa. 2005. *The Birth of Kumāra.* Translated by David Smith. Clay Sanskrit Library. New York: New York University Press, JJC Foundation.

Kāṇe, Pāṇḍuraṅga. 1953. *List of Tīrthas: Being an Off-Print from MM Dr. P. V. Kane's History of Dharmaśāstra.* Vol. 4. Poona: Bhandarkar Oriental Research Institute.

Kar, Sabyasachi. 2007. "Inclusive Growth in Hilly Regions: Priorities for the Uttarakhand Economy." Working paper, Institute of Economic Growth, Delhi.

Kaur, Jagdish. 1985. *Himalayan Pilgrimages and the New Tourism.* New Delhi: Himalayan Books.

Kennedy, Dane Keith. 1996. *The Magic Mountains: Hill Stations and the British Raj.* Berkeley: University of California Press.

Kent, Eliza F. 2010. "Guest Editor's Introduction: Forests of Belonging: The Contested Meaning of Trees and Forests in Indian Hinduism." *Journal for the Study of Religion, Nature and Culture* 4 (2): 129–38. https://doi.org/10.1558/jsrnc.v4i2.129.

———. 2013. *Sacred Groves and Local Gods: Religion and Environmentalism in South India.* New York: Oxford University Press.

Khalid, Amna. 2008. "The Colonial Behemoth: The Sanitary Regulations of Pilgrimage Sites in Northern India, c. 1867–1915." PhD diss., University of Oxford. http://ethos.bl.uk/OrderDetails.do?uin=uk.bl.ethos.496576.

Khandekar, Nivedita, and Radhika Nagrath. 2013. "ASI to Assess Kedarnath Damage, Row over Puja On." *Hindustan Times,* June 28. www.hindustantimes.com/delhi/asi-to-assess-kedarnath-damage-row-over-puja-on/story-CUhtde2dUVtG3KT2j3OHmI.html.

Kinnard, Jacob N. 2014. *Places in Motion: The Fluid Identities of Temples, Images, and Pilgrims.* New York: Oxford University Press.

Knott, Kim. 2009. "From Locality to Location and Back Again: A Spatial Journey in the Study of Religion." *Religion* 39 (2): 154–60.

Kohn, Eduardo. 2013. *How Forests Think toward an Anthropology beyond the Human.* Berkeley: University of California Press.

Kong, Lily. 2010. "Global Shifts, Theoretical Shifts: Changing Geographies of Religion." *Progress in Human Geography* 34 (6): 755–76. https://doi.org/10.1177/0309132510362602.

Kramrisch, Stella. 1981. *The Presence of Śiva.* Princeton, NJ: Princeton University Press.

Krṣṇākumār, ed. 1993. *Kedārakhaṇḍa Purāṇam (mūl saṁskṛt, hindī anuvād evaṁ vistṛt samīkṣā).* Kankhal, Haridvar: Prācya Vidyā Akādamī.

Kukaretī, Viṣṇudatt. 1986. *Nāthpanth: Gaḍhvāl ke pariprekṣya meṅ.* Chaurās-Śrīnagar: Vidyā Niketan Prakāśan.

Kumar, Anup. 2011. *The Making of a Small State: Populist Social Mobilisation and the Hindi Press in the Uttarakhand Movement.* New Delhi: Orient Blackswan.

Kumar, Hari. 2013a. "Relatives Still Searching for 3,000 Missing Pilgrims." *New York Times,* blog *India Ink,* June 28. http://india.blogs.nytimes.com/2013/06/28/relatives-still-searching-for-3000-missing-pilgrims-in-uttarakhand/.

———. 2013b. "When the Himalayas Poured." *New York Times,* blog *India Ink,* June 22. http://india.blogs.nytimes.com/2013/06/22/when-the-himalayas-poured/.

Kumar, Yogesh. 2014. "ASI Peeling Off Ghee from Inside Kedar Shrine." *Times of India,* June 20. http://timesofindia.indiatimes.com/india/ASI-peeling-off-ghee-from-inside-Kedar-shrine/articleshow/36852511.cms.

Kunwar, D. S. 2014. "Snowfall Stalls Work on Trek Route to Kedarnath." *Times of India,* April 13. http://timesofindia.indiatimes.com/india/Snowfall-stalls-work-on-trek-route-to-Kedarnath/articleshow/33710715.cms.

Lakṣmīdhara. 1942. *Tīrthavivecanakaṇḍa.* Edited by K. V. Rangaswamy. Vol. 8. *Kṛyta-Kalpataru.* Baroda: Oriental Institute.

Larkin, Brian. 2013. "The Politics and Poetics of Infrastructure." *Annual Review of Anthropology* 42 (1): 327–43. https://doi.org/10.1146/annurev-anthro-092412-155522.

Latour, Bruno. 1993. *We Have Never Been Modern.* Cambridge, MA: Harvard University Press.

———. 2005. *Reassembling the Social: An Introduction to Actor-Network-Theory.* Oxford: Oxford University Press.

Leavitt, John. 1988. "A Mahabharata Story from the Kumaon Hills." *Himalaya, the Journal of the Association for Nepal and Himalayan Studies* 8 (2): 1–12.

Lehman, Jessica. 2014. "Expecting the Sea: The Nature of Uncertainty on Sri Lanka's East Coast." *Geoforum* 52 (March): 245–56. https://doi.org/10.1016/j.geoforum.2013.05.010.

Leslie, Julia. 1998. "Understanding Basava: History, Hagiography and a Modern Kannada Drama." *Bulletin of the School of Oriental and African Studies, University of London* 61 (2): 228–61.

LeVasseur, Todd. 2015. "'The Earth Is Sui Generis': Destabilizing the Climate of Our Field." *Journal of the American Academy of Religion* 83 (2): 300–319.

Linkenbach, Antje. 2006. "Nature and Politics: The Case of Uttarakhand, North India." In *Ecological Nationalisms: Nature, Livelihoods, and Identities in South Asia,* edited by Gunnel Cederlöf and Kalyanakrishnan Sivaramakrishnan, 151–69. Seattle: University of Washington Press.

———. 2007. *Forest Futures: Global Representations and Ground Realities in the Himalayas.* London: Seagull.

Lochtefeld, James. 2010. *God's Gateway: Identity and Meaning in a Hindu Pilgrimage Place.* New York: Oxford University Press.

Lorenzen, David N. 1972. *The Kāpālikas and Kālāmukhas: Two Lost Śaivite Sects.* Berkeley: University of California Press.

———. 1988. "The Kālāmukha Background to Viraśaivism." In *Studies in Orientology: Essays in Memory of Prof. A. L. Basham,* edited by S. K. Maity, Upendra Thakur, and A. K. Narain, 278–93. Agra: Y. K. Publishers.

Lorimer, Hayden. 2008. "Cultural Geography: Non-representational Conditions and Concerns." *Progress in Human Geography* 32 (4): 551–59. https://doi.org/10.1177/0309132507086882.

Low, Setha M., and Denise Lawrence-Zunigais. 2003. *The Anthropology of Space and Place: Locating Culture.* Malden, MA: Wiley-Blackwell.

Maclean, Kama. 2003. "Making the Colonial State Work for You: The Modern Beginnings of the Ancient Kumbh Mela in Allahabad." *Journal of Asian Studies* 62 (3): 873–905.

———. 2008. *Pilgrimage and Power: The Kumbh Mela in Allahabad, 1765–1954.* New York: Oxford University Press.

Maikhuri, R. K., Vikram S. Negi, L. S. Rawat, and Ajay Maletha. 2014. "Sustainable Development of Disaster-Affected Rural Landscape of Kedar Valley (Uttarakhand) through Simple Technological Interventions." *Current Science* 106 (7): 915.

Majumdar, Ushinor. 2013. "A Ravaged State Pins Its Hope on Tourism." *Tehelka*, December 20. old.tehelka.com/a-ravaged-state-pins-its-hope-on-tourism.

Mallinson, James. 2009. "Nāth Sampradāya." In Jacobsen et al. 2009–15, vol. 1.

Malville, J. McKim, and Baidyanath Saraswati, eds. 2009. *Pilgrimage: Sacred Landscapes and Self-Organized Complexity.* New Delhi: Indira Gandhi National Centre for the Arts; D.K. Printworld.

Mamgain, Rajendra P. 2004. "Employment, Migration and Livelihoods in the Hill Economy of Uttaranchal." PhD diss., Jawaharlal Nehru University, New Delhi. Reproduced as MPRA Paper No. 32303, posted July 19, 2011, 13:35 UTC, Munich Personal RePEc Archive. http://mpra.ub.uni-muenchen.de/32303/.

———. 2008. "Growth, Poverty and Employment in Uttarakhand." *Journal of Labour and Development* 13 (2). www.ihdindia.org/%5C/Working%20Ppaers/2010-2005/pdf%20files/39-%20RP%20Mamgain.pdf.

Mathur, Nayanika. 2015a. "'It's a Conspiracy Theory and Climate Change': Of Beastly Encounters and Cervine Disappearances in Himalayan India." *HAU: Journal of Ethnographic Theory* 5 (1): 87–111.

———. 2015b. "A 'Remote' Town in the Indian Himalaya." *Modern Asian Studies* 49 (2): 365–92. https://doi.org/10.1017/S0026749X1300053X.

Mawdsley, Emma. 1999. "A New Himalayan State in India: Popular Perceptions of Regionalism, Politics, and Development." *Mountain Research and Development* 19 (2): 101–12.

———. 2005. "The Abuse of Religion and Ecology: The Vishva Hindu Parishad and Tehri Dam." *Worldviews: Global Religions, Culture and Ecology* 9 (1): 1–24.

Mazoomdaar, Jay. 2013. "How Uttarakhand Dug Its Grave." Mazoomdar (blog), June 28. http://mazoomdaar.blogspot.com/2013/06/how-uttarakhand-dug-its-grave.html

McCarter, Elliott Craver. 2013. "Kurukshetra: Bending the Narrative into Place." PhD diss., University of Texas, Austin. http://repositories.lib.utexas.edu/handle/2152/21942.

McGregor, R., ed. 1993. *The Oxford Hindi-English Dictionary.* Oxford: Oxford University Press.

Mehta, M., Z. Majeed, D. P. Dobhal, and P. Srivastava. 2012. "Geomorphological Evidences of Post-LGM Glacial Advancements in the Himalaya: A Study from Chorabari Glacier, Garhwal Himalaya, India." *Journal of Earth System Science* 121 (1): 149–63.

Michael, R. Blake. 1983. "Foundation Myths of the Two Denominations of Vīraśaivism: Viraktas and Gurusthalins." *Journal of Asian Studies* 42 (2): 309–22. https://doi. org/10.2307/2055116.

Mihir, Srivastava, and Raul Irani. 2013. "Anatomy of a Disaster." *OPEN,* June 27. www. openthemagazine.com/article/nation/anatomy-of-a-disaster.

Mishra, Saurabh. 2011. *Pilgrimage, Politics, and Pestilence: The Haj from the Indian Subcontinent, 1860–1920.* Oxford: Oxford University Press.

Moore, Donald. 2003. "Beyond Blackmail: Multivalent Modernities and the Cultural Politics of Development in India." In *Regional Modernities: The Cultural Politics of Development in India,* edited by Arun Agrawal and Kalyanakrishnan Sivaramakrishnan, 165–214. Stanford, CA: Stanford University Press.

Mudaliar, Chandra. 1976. *State and Religious Endowments in Madras.* Madras: University of Madras.

Mukherjea, B. K., Saiyid Fazl Ali, and S. K. Das. 1952. Nar Hari Sastri and Ors. Vs. Shri Badrinath Temple Committee, Supreme Court of India. May 9. https://indiankanoon. org/doc/801837/.

Mukherjee, Pampa. 2012. "Globality and Imagined Futures: A Brief History of Uttarakhand's Development Dreams." In *Facing Globality: Politics of Resistance, Relocation, and Reinvention in India,* edited by Bhupinder Brar and Pampa Mukherjee, 201–33. New Delhi: Oxford University Press.

Muller-Ortega, Paul Eduardo. 1989. *The Triadic Heart of Siva: Kaula Tantricism of Abhinavagupta in the Non-dual Shaivism of Kashmir.* Albany: State University of New York Press.

Nagarajan, Vijaya Rettakudi. 1998. "The Earth as Goddess Bhu Devi: Towards a Theory of 'Embedded Ecologies' in Folk Hinduism." In Nelson 1998, 269–96.

Naithānī, Śivprasad. 2006. *Uttarākhaṇḍ ke tīrth evaṃ mandir: Himālaya, yātrā aur paryaṭan sahit.* 3rd ed. Srīnagar, Gaṛhwāl: Pavetrī Prakāśan.

Nandy, Ashish. 1980. *At the Edge of Psychology: Essays in Politics and Culture.* Delhi: Oxford University Press.

Nath, Vijay. 2001. *Purāṇas and Acculturation: A Historico-Anthropological Perspective.* New Delhi: Munshiram Manoharlal.

———. 2009. *The Puranic World: Environment, Gender, Ritual, and Myth.* New Delhi: Manohar.

Nautiyal, Shivanand, ed. 1994. *Kedārakhaṇḍam—Skandapurāṇāntargata (Hindī anuvād sahit).* Translated by Shivanand Nautiyal. Ilahabad: Hindi Sahitya Sammelan.

Negi, S. S. 2001. "Back and the Beyond: Races, Dynasties, Rulers, Socio-Cultural Setup." In *Garhwal Himalaya: Nature, Culture and Society,* edited by O. P. Kandari and O. P. Gusain, 3–34. Srinagar, Garwhal: TransMedia.

Nelson, Lance E., ed. 1998. *Purifying the Earthly Body of God: Religion and Ecology in Hindu India.* Albany: State University of New York Press.

Neumann, Roderick P. 2009. "Political Ecology: Theorizing Scale." *Progress in Human Geography* 33 (3): 398–406. http://dx.doi.org/10.1177/0309132508096353.

Nivedita, Sister. 1928. *Kedar Nath and Badri Narayan: (A Pilgrim's Diary).* 2nd ed. Calcutta: Udbodhan Office.

Oakley, E. Sherman. 1905. *Holy Himalaya: The Religion, Traditions, and Scenery of a Himalayan Province (Kumaon and Garhwal).* Edinburgh: Oliphant Anderson and Ferrier.

Oliver-Smith, Anthony. 1999a. "Peru's Five-Hundred-Year Earthquake: Vulnerability in Historical Context." In Oliver-Smith and Hoffman 1999, 74–88.

———. 1999b. "What Is a Disaster? Anthropological Perspectives on a Persistent Question." In Oliver-Smith and Hoffman 1999, 18–34.

Oliver-Smith, Anthony, and Susanna M. Hoffman, eds. 1999. *The Angry Earth: Disaster in Anthropological Perspective*, edited by Anthony Oliver-Smith and Susanna M. Hoffman. New York: Routledge.

Outlook India Staff. 2014. "44 Skeleton Remains Found in Kedar Valley." *Outlook India*, June 8. www.outlookindia.com/news/article/44-Skeleton-Remains-Found-in-Kedar-Valley/845208.

Padma, Sree. 2013. *Vicissitudes of the Goddess: Reconstructions of the Gramadevata in India's Religious Traditions.* New York: Oxford University Press.

Padmanabhan, Chitra. 2013. "When the Ganga Descends." *Hindu,* June 29. www.thehindu.com/opinion/op-ed/when-the-ganga-descends/article4857510.ece.

Padumā, Rām Nārāyaṇasinhajūdeva Bahādur, and J. Hajārībāg, trans. 1907. *Śrīkedārakalpa.* Bombay: Khemarāja Śrīkṛṣṇadās (Śrī Veṅkaṭeśvar Press).

Paljor, Karma. 2013. "Rahul Gandhi Visits Flood-Hit Uttarakhand, to Monitor Relief Work." *IBNLive,* June 24. http://ibnlive.in.com/news/rahul-gandhi-visits-floodhit-uttarakhand-to-monitor-relief-work/401502-37-64.html.

Pande, Govind Chandra. 1994. *Life and Thought of Śaṅkarācārya.* Delhi: Motilal Banarsidass.

Pāṇḍey, Dhīrendra Nāth. 2014. *Āpdā kā kafan: Pīṛit yuvatī kī saṁgharṣ gāthā.* Dehra Dun: Jyoti.

Pathak, Shekhar. 2013. "Ye vikās kī hoḍ hai yā tabāhī ko dāvat?" *BBC Hindi,* June 20. www.bbc.co.uk/hindi/india/2013/06/130619_uttarakhand_kedarnath_disaster_vs.shtml.

Pathak, Vishwambhar Sharan. 1987. *Smārta Religious Tradition: Being a Study of the Epigraphic Data on the Smārta Religious Tradition of Northern India, c. 600 A.D. to c. 1200 A.D.* Meerut, India: Kusumanjali Prakashan.

Patton, Laurie L., and Wendy Doniger, eds. 1996. *Myth and Method.* Charlottesville: University of Virginia Press.

Peer, Basharat. 2013. "Flood Toll Reaches 1,000 in India as Thousands More Await Rescue." *New York Times,* June 22. www.nytimes.com/2013/06/23/world/asia/flooding-kills-hundreds-in-northern-india.html.

Petley, Dave. 2013. "New High Resolution Images of Kedarnath: The Cause of the Debris Flow Disaster Is Now Clear." *Landslide Blog,* June 27. http://blogs.agu.org/landslideblog/2013/06/27/new-high-resolution-images-of-kedarnath-the-cause-of-the-debris-flow-disaster-is-now-clear/.

Pinch, William R. 2006. *Warrior Ascetics and Indian Empires.* Cambridge: Cambridge University Press.

Pinkney, Andrea Marion. 2013a. "An Ever-Present History in the Land of the Gods: Modern Māhātmya Writing on Uttarakhand." *International Journal of Hindu Studies* 17 (3): 229–60. https://doi.org/10.1007/s11407-014-9142-1.

———. 2013b. "Prasada, the Gracious Gift, in Contemporary and Classical South Asia." *Journal of the American Academy of Religion Journal of the American Academy of Religion* 81 (3): 734–56.

Pintchman, Tracy. 1994. *The Rise of the Goddess in the Hindu Tradition.* Albany: State University of New York Press.

Podār, Hanumān Prasād, Cimmanlāl Goswāmi, and M. E. Shāstri, eds. 1957. *Kalyāṇ Tīrthāṅk.* Gorakhpur: Gīta Press.

Polit, Karin M. 2008. "Performing Heritage through Rituals, the Case of Uttarakhand." In *Religion, Ritual, Theatre,* edited by Bent Fleming Nielsen, Bent Holm, and Karen Vedel, 107–22. Frankfurt: Peter Lang.

———. 2012. *Women of Honour: Gender and Agency among Dalit Women in the Central Himalayas.* New Delhi: Orient Blackswan.

Prasad, Leela. 2006. *Poetics of Conduct: Oral Narrative and Moral Being in a South Indian Town.* New York: Columbia University Press.

Presler, Franklin A. 1987. *Religion under Bureaucracy: Policy and Administration for Hindu Temples in South India.* Cambridge: Cambridge University Press.

Press Information Bureau, Government of India. 2015. "PRASAD Scheme of Ministry of Tourism." April 27. http://pib.nic.in/newsite/mbErel.aspx?relid=119771.

Press Trust of India. 2013a. "Bahuguna Turns Down Modi's Offer to Reconstruct Kedarnath Shrine." *Hindu,* June 30. www.thehindu.com/news/national/bahuguna-turns-down-modis-offer-to-reconstruct-kedarnath-shrine/article4865991.ece.

———. 2013b. "Epidemic Fear Spurs Rush to Complete Last Rites of Uttarakhand Victims." *Hindu,* June 25. www.thehindu.com/news/national/epidemic-fear-spurs-rush-to-complete-last-rites-of-uttarakhand-victims/article4849164.ece.

———. 2013c. "Modi Trashes Shinde's Request; Tours Flood-Hit Uttarakhand." *Rediff News,* June 22. www.rediff.com/news/report/modi-trashes-shindes-request-tours-flood-hit-uttarakhand/20130622.htm.

———. 2013d. "Pilgrims Return Home with Uttarakhand Horror Tales." *Hindustan Times,* June 21. www.hindustantimes.com/india/pilgrims-return-home-with-uttarakhand-horror-tales/story-4jPP33GlhhmFMlo901CmrJ.html.

———. 2013e. "Prayers to Resume at Kedarnath Shrine from September 11: Chief Minister Vijay Bahuguna." NDTV, September 1. www.ndtv.com/article/india/prayers-to-resume-at-kedarnath-shrine-from-september-11-chief-minister-vijay-bahuguna-413012.

———. 2013f. "Ration Meant for Flood-Hit U'khand Rotting in Godowns." *Outlook India,* December 30. www.outlookindia.com/news/article/Ration-Meant-for-FloodHit-Ukhand-Rotting-in-Godowns/822943.

———. 2013g. "Roads Connecting Kedarnath Shrine to Be Restored by September: Oscar Fernandes." *Times of India,* July 10. http://articles.timesofindia.indiatimes.com/2013-07-10/india/40491682_1_oscar-fernandes-kedarnath-shrine-worst-hit-rudraprayag.

———. 2013h. "Uttarakhand Disaster: Army Opening New Route to Kedarnath." NDTV. com, July 11. www.ndtv.com/article/india/uttarakhand-disaster-army-opening-new-route-to-kedarnath-390789.

———. 2013i. "Uttarakhand: Pilgrim Clung to Bell at Kedarnath Temple for Nine Hours, Claims Family." NDTV.com, June 24. www.ndtv.com/article/india/uttarakhand-pilgrim-clung-to-bell-at-kedarnath-temple-for-nine-hours-claims-family-383479.

———. 2013j. "Uttarakhand: Search and Rescue Operations in Kedarnath Over." *Times of India*, June 25. http://timesofindia.indiatimes.com/india/Uttarakhand-Search-and-rescue-operations-in-Kedarnath-over/articleshow/20756994.cms.

———. 2013k. "Uttarkhand Team Visits Odisha." *Outlook India*, December 19. www.outlookindia.com/news/article/Uttarkhand-Team-Visits-Odisha/821612.

———. 2014a. "Rahul Gandhi Meets Kin of Uttarakhand Tragedy Victims." *Outlook India*, January 26. www.outlookindia.com/news/article/Rahul-Gandhi-Meets-Kin-of-Uttarakhand-Tragedy-Victims/826287.

———. 2014b. "Rawat Orders Completion of Gaurikund-Kedarnath Road before Char Dham." *Zee News*, February 7. http://zeenews.india.com/news/uttarakhand/rawat-orders-completion-of-gaurikund-kedarnath-road-before-char-dham_909883.html.

———. 2014c. "U'khand: BJP Slams Govt for Failure to Act on Directions." *Outlook India*, January 21. www.outlookindia.com/newswire/story/ukhand-bjp-slams-govt-for-failure-to-act-on-directions/825714.

———. 2014d. "Uttarakhand Chief Minister Vijay Bahuguna Quits." *Outlook India*, January 31. www.outlookindia.com/news/article/Uttarakhand-Chief-Minister-Vijay-Bahuguna-Quits/827019.

Prudham, Scott, and Nik Heynen. 2011. "Introduction: Uneven Development 25 Years On: Space, Nature and the Geographies of Capitalism." *New Political Economy* 16 (2): 223–32. https://doi.org/10.1080/13563467.2011.542806.

Purohit, Data Ram, Richa Negi, and Poornanand Negi. 1995. "Perception of Bhutas in Garhwal." In *Primal Elements: The Oral Tradition,* edited by Baidyanath Saraswati, 1:55–69. Prakriti: The Integral Vision. New Delhi: Indira Gandhi National Centre for the Arts.

Ramanujan, A. K., ed. and trans. 1967. *The Interior Landscape; Love Poems from a Classical Tamil Anthology.* UNESCO Collection of Representative Works. Bloomington: Indiana University Press.

Rangan, Haripriya. 1995. "Contested Boundaries: State Policies, Forest Classifications, and Deforestation in the Garhwal Himalayas." *Antipode* 27 (4): 343–62. https://doi.org/10.1111/j.1467-8330.1995.tb00284.x.

———. 2000. *Of Myths and Movements: Rewriting Chipko into Himalayan History.* London: Verso.

———. 2004. "From Chipko to Uttaranchal: The Environment of Protest and Development in the Indian Himalaya." In *Liberation Ecologies: Environment, Development, Social Movements,* edited by Richard Peet, 205–26. London: Routledge.

Rao, K. L. Sheshagiri. 2000. "The Five Great Elements (Pañcamahābhūta): An Ecological Perspective." In *Hinduism and Ecology: The Intersection of Earth, Sky, and Water,* edited by Christopher Key Chapple and Mary Evelyn Tucker, 23–38. Cambridge, MA: Harvard University Press for the Center for the Study of World Religions, Harvard Divinity School.

Rautela, Piyoosh. 2013. "Lessons Learnt from the Deluge of Kedarnath, Uttarakhand, India." *Asian Journal of Environment and Disaster Management* 5 (2): 43–51. doi: https://doi.org/10.3850/S1793924013002824.

Rautela, Vikram. 2014. "Skeletons Found in Flood-Ravaged Rambara." *Times of India*, April 1. http://timesofindia.indiatimes.com/india/Skeletons-found-in-flood-ravaged-Rambara/articleshow/33047643.cms.

Rāvat, Śivrāj Siṃh. 2006. *Kedār Himālay aur Pañc Kedār*. Dehra Dun: Vinsar.

Rawat, Ajay Singh. 2002. *Garhwal Himalaya: A Study in Historical Perspective*. New Delhi: Indus.

Ray, Satyajit. 2000. *The Complete Adventures of Feluda*. Translated by Gopa Majumdar. Vol. 2. New Delhi: Penguin Books India.

Reader, Ian. 2014. *Pilgrimage in the Marketplace*. New York: Routledge.

Reddy, Prabhavati C. 2014. *Hindu Pilgrimage: Shifting Patterns of Worldview of Shrī Shailam in South India*. Routledge Hindu Studies. London: Routledge.

Rediff. 2014. "BJP Leader Rescued from Kedarnath Says 20,000 Dead." *Rediff News*, June 20. www.rediff.com/news/report/bjp-leader-rescued-from-kedarnath-says-20000-dead/20130620.htm.

Richardson, Shaun D., and John M. Reynolds. 2000. "An Overview of Glacial Hazards in the Himalayas." *Quaternary International* 65/66: 31–47.

Ritter, Valerie. 2011. *Kāma's Flowers: Nature in Hindi Poetry and Criticism, 1885–1925*. Albany: State University of New York Press.

Rizvi, Janet. 1981. "The Fragile Himalaya: A Review Article." *India International Centre Quarterly* 8 (2): 173–82.

Rose, Mitch. 2006. "Gathering 'Dreams of Presence': A Project for the Cultural Landscape." *Environment and Planning. D: Society and Space* 24 (4): 537–54.

Rose, Mitch, and John Wylie. 2006. "Animating Landscape." *Environment and Planning D: Society and Space* 24 (4): 475–79.

Roy, Srirupa. 2006. "'A Symbol of Freedom': The Indian Flag and the Transformations of Nationalism, 1906–2002." *Journal of Asian Studies* 65 (3): 495–527.

Saklani, Dinesh Prasad. 1998. *Ancient Communities of the Himalaya*. New Delhi: Indus.

Samuel, Geoffrey. 2008. *The Origins of Yoga and Tantra: Indic Religions to the Thirteenth Century*. Cambridge: Cambridge University Press.

Sanderson, Alexis. 1988. "Śaivism and the Tantric Traditions." In *The World's Religions*, edited by S. Sutherland, L. Houlden, P. Clarke, and F. Hardy, 660–704. London: Routledge and Kegan Paul.

———. 2009. "The Śaiva Age: The Rise and Dominance of Śaivism during the Early Medieval Period." In *Genesis and Development of Tantrism*, edited by Shingo Einoo, 41–350. Tokyo: Institute of Oriental Culture, University of Tokyo.

———. 2013. "The Impact of Inscriptions on the Interpretation of Early Śaiva Literature." *Indo-Iranian Journal* 56 (3–4): 211–44.

Sāṅkṛtyāyana, Rāhula. 1953 *Himālaya-paricaya*. Ilāhābāda. Ilāhābāda Lā Jarnala Preṣa.

Sarbadhikary, Sukanya. 2015. *The Place of Devotion*. Berkeley: University of California Press. https://doi.org/10.1525/luminos.2.

Sati, Vishwambhar Prasad. 2013. "Tourism Practices and Approaches for Its Development in the Uttarakhand Himalaya, India." *Journal of Tourism Challenges and Trends* 6 (1): 97–111.

Saunders, T. J. 1844. *Notes of Wanderings in the Himmala, Containing Descriptions of Some of the Grandest Scenery of the Snowy Range; among Others of Nainee Tal*. Agra: T. W. Brown. http://hdl.handle.net/2027/uc1.$b195936.

Sax, William S. 1991. *Mountain Goddess: Gender and Politics in a Himalayan Pilgrimage.* New York: Oxford University Press.

———. 2000. "Residence and Ritual in the Garhwal Himalaya." In *Himalaya: Past and Present,* edited by Maheshwar P. Joshi, Allen C. Fanger, and Charles W. Brown, 4:79–114. Almora, India: Shree Almora Book Depot.

———. 2002. *Dancing the Self: Personhood and Performance in the Pāṇḍav Līlā of Garhwal.* Oxford: Oxford University Press.

———. 2006. "A Divine Identity Crisis." In *Ritual and Identity: Performative Practices as Effective Transformations of Social Reality,* edited by Klaus-Peter Köpping, Bernhard Leistle, and Michael Rudolph, 8:101–28. Performances. Berlin: LIT Verlag.

———. 2009. *God of Justice: Ritual Healing and Social Justice in the Central Himalayas.* New York: Oxford University Press.

———. 2011. "Religion, Rituals, and Symbols of Belonging: The Case of Uttarakhand." In *The Politics of Belonging in the Himalayas: Local Attachments and Boundary Dynamics,* edited by Joanna Pfaff-Czarnecka and Gérard Toffin, 167–81. New Delhi: Sage Publications.

Schlehe, J. 2010. "Anthropology of Religion: Disasters and the Representations of Tradition and Modernity." *Religion* 40 (2): 112–20. https://doi.org/10.1016/j.religion.2009.12.004.

Seeman, Don. 2009. *One People, One Blood: Ethiopian-Israelis and the Return to Judaism.* New Brunswick NJ: Rutgers University Press.

Selby, Martha Ann, and Indira Viswanathan Peterson, eds. 2008. *Tamil Geographies: Cultural Constructions of Space and Place in South India.* SUNY Series in Hindu Studies. Albany: State University of New York Press.

Sharma, Mahesh. 2009. *Western Himalayan Temple Records: State, Pilgrimage, Ritual and Legality in Chambā.* Leiden: Brill.

Sharma, Pitamber. 2000. "Badrinath: Managing Pilgrimage Tourism in a Unique Environment." In *Tourism as Development: Case Studies from the Himalaya,* edited by Pitamber Sharma, 147–66. Lalitpur: Himal Books.

Sharma, Seema. 2013. "Revival Plan for Tourism and Its Dependents in Kedarnath Valley." *Times of India,* December 23. http://timesofindia.indiatimes.com/india/Revival-plan-for-tourism-and-its-dependents-in-Kedarnath-valley/articleshow/27764433.cms.

———. 2014. "Channel Mandakini River to Its Original Course: GSI." *Times of India,* January 14. http://timesofindia.indiatimes.com/india/Channel-Mandakini-river-to-its-original-course-GSI/articleshow/28721142.cms.

Shastri, J. L., ed. 1970. *The Siva-Purana.* 4 vols. Delhi: Motilal Banarsidass.

Sherma, Rita DasGupta. 1998. "Sacred Immanence: Reflections of Ecofeminism in Hindu Tantra." In Nelson 1998, 89–132.

Shiva, Vandana. 1988. *Staying Alive: Women, Ecology, and Development.* London: Zed Books.

Shobhi, Prithvi Datta Chandra. 2005. "Pre-modern Communities and Modern Histories: Narrating Vīraśaiva and Lingayat Selves." PhD diss., University of Chicago.

Shree Badarinath-Shri Kedarnath Temples Committee. n.d. "Number of Pilgrims." Accessed May 23, 2018. www.badarikedar.org/management/Number-Of-Pilgrims.

Shulman, David. 1976. "The Murderous Bride: Tamil Versions of the Myth of Devī and the Buffalo-Demon." *History of Religions* 16 (2): 120–46.

———. 1980. *Tamil Temple Myths: Sacrifice and Divine Marriage in the South Indian Saiva Tradition.* Princeton, NJ: Princeton University Press.

Simpson, Edward. 2014. *The Political Biography of an Earthquake: Aftermath and Amnesia in Gujarat, India*. New York: Oxford University Press.

———. 2016. "Is Anthropology Legal? Earthquakes, Blitzkrieg, and Ethical Futures." *Focaal* 2016 (74). doi: https://doi.org/10.3167/fcl.2016.740109.

Simpson, Edward, and Malathi De Alwis. 2008. "Remembering Natural Disaster: Politics and Culture of Memorials in Gujarat and Sri Lanka." *Anthropology Today* 24 (4): 6–12. doi: https://doi.org/10.1111/j.1467-8322.2008.00599.x

Singh, Kautilya. 2014. "Kedarnath Temple to Reopen on May 4, Challenges Remain for Govt." *Times of India*, February 28. http://timesofindia.indiatimes.com/india/Kedarnath-temple-to-reopen-on-May-4-challenges-remain-for-govt/articleshow/31127398.cms.

Singh, Vijay. 2013. "Kedārnāth āpdā: Apne ghar lauṭe saikṛoṁ lāpatā log." *Amar Ujala*, December 4, Dehra Dun edition. www.dehradun.amarujala.com/news/city-hulchul-dun/kedarnath-disaster-missing-people-returned-home/.

Sircar, Dines Chandra. 1971. *Studies in the Geography of Ancient and Medieval India*. 2nd rev. and enl. ed. Delhi: Motilal Banarsidass.

Smith, Jonathan Z. 1987. *To Take Place: Toward Theory in Ritual*. Chicago Studies in the History of Judaism. Chicago: University of Chicago Press.

Smith, Neil. 2008. *Uneven Development: Nature, Capital, and the Production of Space*. 3rd ed. Athens: University of Georgia Press.

———. 2011. "Uneven Development Redux." *New Political Economy* 16 (2): 261–65. https://doi.org/10.1080/13563467.2011.542804.

Smith, Travis LaMar. 2007. "The Sacred Center and Its Peripheries: Śaivism and the Vārāṇasī Sthala-Purāṇas." PhD diss., Columbia University.

Snodgrass, Jeffrey G., and Kristina Tiedje. 2008. "Indigenous Nature Reverence and Conservation: Seven Ways of Transcending an Unnecessary Dichotomy." *Journal for the Study of Religion, Nature and Culture* 2 (1): 6–29. https://doi.org/10.1558/jsrnc.v2i1.6.

Sridharan, E. 2004. "The Growth and Sectoral Composition of India's Middle Class: Its Impact on the Politics of Economic Liberalization." *India Review* 3 (4): 405–28. https://doi.org/10.1080/14736480490895769.

Srinivas, Mysore Narasimhachar. 1952. *Religion and Society among the Coorgs of South India*. Oxford: Clarendon Press.

Srinivasan, Doris. 1997. *Many Heads, Arms, and Eyes: Origin, Meaning, and Form of Multiplicity in Indian Art*. Leiden: Brill.

Śrī Śivamahāpurāṇa. 2004. Mumbai: Khemarāja Śrīkṛṣṇadās (Śrī Veṅkaṭeśvar Press).

Śrīskandamahāpurāṇāntargatakedārakhaṇḍaḥ. 1906. Mumbai: Khemarāja Śrīkṛṣṇadās / Śrī Veṅkaṭeśvar Press.

State Unit Uttarakhand. 2016. "Preliminary Assessment of Kedarnath Area after the Floods of 16th and 17th June 2013." Geological Survey of India, Dehra Dun. Accessed July 7 (no longer accessible). www.portal.gsi.gov.in/gsiImages/information/gsi%20team%20at%20kedarnath.pdf.

Steinberg, Theodore. 2000. *Acts of God: The Unnatural History of Natural Disaster in America*. New York: Oxford University Press.

Sutherland, Gail Hinich. 1991. *The Disguises of the Demon: The Development of the Yakṣa in Hinduism and Buddhism*. Albany: State University of New York Press.

Sutherland, Peter. 1998. "Travelling Gods and Government by Deity: An Ethnohistory of Power, Representation and Agency in West Himalayan Polity." PhD diss., Oxford University.

———. 2006. "T(r)Opologies of Rule (Raj): Ritual Sovereignty and Theistic Subjection." *European Bulletin of Himalayan Research* 29–30:82–119.

Tandon, Rajesh J. 2007. Shri Badri Nath Kedarnath Temples Committee vs. Sri Mahadeo Prasad and ors., High Court of Uttarakhand. Photocopied document in possession of author.

Thal, Sarah. 2005. *Rearranging the Landscape of the Gods: The Politics of a Pilgrimage Site in Japan, 1573–1912.* Chicago: University of Chicago Press.

Thapliyal, A. P., Jina Mandal, K. Lakshmanan, P. V. S. Rawat, Harish Bahuguna, and S. K. Tripathi. 2013. "Preliminary Slope Stability Assessment of the Recent Disaster Affected Areas of Rudraprayag District, Uttarakhand." Category-B Open File Report LHZ/NR/UPUK/2013/052. Geological Survey of India.

Thien, Deborah. 2005. "After or beyond Feeling? A Consideration of Affect and Emotion in Geography." *Area* 37 (4): 450–54. https://doi.org/10.1111/j.1475-4762.2005.00643a.x.

"360 Teerth Purohit." n.d. Accessed September 18, 2014. https://360tp.wordpress.com/.

Thrift, Nigel. 2004. "Intensities of Feeling: Towards a Spatial Politics of Affect." *Geografiska Annaler: Series B, Human Geography* 86 (1): 57–78. https://doi.org/10.1111/j.0435-3684.2004.00154.x.

Tillotson, G. H. R. 1990. "The Indian Picturesque: Images of India in British Landscape Painting, 1780–1880." In *The Raj: India and the British, 1600–1947,* edited by C. A. Bayly, 141–51. London: National Portrait Gallery Publications.

Tolia-Kelly, D. P. 2007. "Fear in Paradise: The Affective Registers of the English Lake District Landscape Re-visited." *Senses and Society* 2 (3): 329–51.

Tomalin, Emma. 2004. "Bio-divinity and Biodiversity: Perspectives on Religion and Environmental Conservation in India." *Numen* 51 (3): 265–95. https://doi.org/10.1163/1568527041945481.

Törzsök, Judit. 2011. "Kāpālikas." In Jacobsen et al. 2009–15, vol. 3.

Traill, George William. 1823. "Report on the Districts of Kumaon and Garhwal by the Commissioner, George William Traill (with Associated Correspondence)." Records of the Board of Commissioners for the Affairs of India, 1784–1858. IOR/F/4/828/21951, India Office Library, London.

Tripathi, Purnima S. 2013. "Himalayan Tragedy." *Frontline,* July 13, 2013. www.frontline.in/the-nation/himalayan-tragedy/article4840776.ece.

Tsing, Anna Lowenhaupt. 2015. *Friction: An Ethnography of Global Connection.* Princeton, NJ: Princeton University Press.

Tucker, Catherine M, ed. 2012. *Nature, Science, and Religion: Intersections Shaping Society and the Environment.* Santa Fe, NM: School for Advanced Research Press.

Tweed, Thomas A. 2006. *Crossing and Dwelling: A Theory of Religion.* Cambridge, MA: Harvard University Press.

UN Office for Disaster Risk Reduction. 2007. "Hyogo Framework for Action (HFA)." https://www.unisdr.org/we/coordinate/hfa.

Upadhyay, Kavita. 2013a. "As Religious Tourism Declines, Ukhimath Residents Face a New Challenge." *Hindu,* December 8. www.thehindu.com/news/national/other-states/as-religious-tourism-declines-ukhimath-residents-face-a-new-challenge/article5437177.ece.

————. 2013b. "82,000 Evacuated So Far; 22,000 Still Stranded in Uttarakhand." *Hindu*, June 22. www.thehindu.com/news/national/82000-evacuated-so-far-22000-still-stranded-in-uttarakhand/article4840119.ece.

————. 2013c. "15 More Bodies Found in Kedarnath." *Hindu*, October 29. www.thehindu.com/news/national/other-states/15-more-bodies-found-in-kedarnath/article5285766.ece.

————. 2013d. "Villagers Live in Cow Sheds, Tents Even as Severe Winters Approach." *Hindu*, December 30. www.thehindu.com/news/national/other-states/villagers-live-in-cow-sheds-tents-even-as-severe-winters-approach/article5519215.ece.

————. 2014a. "Char Dham Yatra to 'Be Regulated Not Restricted.'" *Hindu*, March 16. www.thehindu.com/news/national/other-states/char-dham-yatra-to-be-regulated-not-restricted/article5792779.ece.

————. 2014b. "Rawat Prescribes an Image Change for Uttarakhand." *Hindu*, February 10. www.thehindu.com/todays-paper/tp-miscellaneous/tp-others/rawat-prescribes-an-image-change-for-uttarakhand/article5671997.ece.

Valdiya, K. S. 2014. "Damming Rivers in the Tectonically Resurgent Uttarakhand Himalaya." *Current Science* 106 (12): 1658–68.

Van der Veer, Peter. 1997. *Gods on Earth: Religious Experience and Identity in Ayodhya*. Delhi: Oxford University Press.

————. 2014. *The Modern Spirit of Asia: The Spiritual and the Secular in China and India*. Princeton, NJ: Princeton University Press.

Vásquez, Manuel A. 2011. *More Than Belief: A Materialist Theory of Religion*. Oxford: Oxford University Press.

Vassilkov, Yaroslav. 2002. "Indian Practice of Pilgrimage and the Growth of the Mahabharata in the Light of New Epigraphical Sources." In *Stages and Transitions: Temporal and Historical Frameworks in Epic and Purāṇic Literature*, edited by Mary Brockington, 133–53. Proceedings of the Second Dubrovnik International Conference on the Sanskrit Epics and Puranas. Zagreb: Croatian Academy of Arts and Sciences.

Veldman, Robin Globus, Andrew Szasz, and Randolph Haluza-DeLay. 2012. "Introduction: Climate Change and Religion—a Review of Existing Research." *Journal for the Study of Religion, Nature and Culture* 6 (3): 255–75. https://doi.org/10.1558/jsrnc.v6i3.255.

Vembu, Venky. 2014. "Safety First." *Outlook India*, April 1. http://archive.outlooktraveller.com/article.aspx?289897.

Vīramitrodayaḥ. 1917. *Vīramitrodayaḥ tīrtha prakāśa*. Edited by Viṣṇuprasāda and Mitra Miśra. Chowkhamba Sanskrit Series. Benares: Chowkhamba Sanskrit Series Office.

Viśālmaṇi Śarmā Upādhyāy. 1952. *Kedāra kalpa*. Nārayaṇ Koṭi, Gaṛhvāl: Viśālmaṇi Śarmā Upādhyāya.

Vyāsa. 2007. *Maharṣivyāsapraṇītaḥ Skandamahāpurāṇāntargataḥ Kedārakhaṇḍaḥ: Ratnaprabhābhāṣāvyākhyāsahita*. Translated by Vrajavallia Bhaṭṭācārya. Lucknow: Maheśvaraprasāda Śrīvāstava; Vitaraka Caukhambā Saṃskṛta Pratiṣṭhāna.

Wadley, Susan Snow. 1975. *Shakti: Power in the Conceptual Structure of Karimpur Religion*. Series in Social, Cultural, and Linguistic Anthropology, no. 2. Chicago: Department of Anthropology, University of Chicago.

Walton, H. G. [1910] 1989. *British Garhwal: A Gazetteer*. Dehra Dun: Natraj.

Warf, Barney, and Santa Arias. 2009. *The Spatial Turn: Interdisciplinary Perspectives*. London: Routledge.

Wasson, R. J., N. Juyal, M. Jaiswal, M. McCulloch, M. M. Sarin, V. Jain, P. Srivastava, et al. 2008. "The Mountain-Lowland Debate: Deforestation and Sediment Transport in the Upper Ganga Catchment." *Journal of Environmental Management* 88 (1): 53–61. https://doi.org/10.1016/j.jenvman.2007.01.046.

Wasson, R. J., Y. P. Sundriyal, Shipra Chaudhary, Manoj K. Jaiswal, P. Morthekai, S. P. Sati, and Navin Juyal. 2013. "A 1000-Year History of Large Floods in the Upper Ganga Catchment, Central Himalaya, India." *Quaternary Science Reviews* 77 (October): 156–66. https://doi.org/10.1016/j.quascirev.2013.07.022.

Whatmore, Sarah. 2006. "Materialist Returns: Practising Cultural Geography in and for a More-Than-Human World." *Cultural Geographies* 13 (4): 600–609.

White, David Gordon. 1996. *The Alchemical Body: Siddha Traditions in Medieval India.* Chicago: University of Chicago Press.

———. 2009. "Bhairava." In Jacobsen et al. 2009–15, vol. 1.

———. 2015. "Tantra." In Jacobsen et al. 2009–15, vol. 6.

Whitmore, Luke. 2012. "The Challenges of Representing Shiva: Image, Place, and Divine Form in the Himalayan Hindu Shrine of Kedarnath." *Material Religion: The Journal of Objects, Art and Belief* 8 (2): 215–41.

———. 2016. "In the Mountains of Radical Juxtaposition: Kedarnath at the Beginning of a New Millennium." In *Religion and Modernity in the Himalaya,* edited by Jessica Vantine Birkenholtz and Megan Adamson Sijapati, 25–42. New York: Routledge.

———. 2018. "Changes in Ritual Practice at the Himalayan Hindu Shrine of Kedarnath." In *Ritual Innovation: Strategic Interventions in South Asian Religion,* edited by Brian K. Pennington and Amy L. Allocco, 71–90. Albany: State University of New York Press.

Wikan, Unni. 1991. "Toward an Experience-Near Anthropology." *Cultural Anthropology* 6 (3): 285–305.

Willen, Sarah S. 2007. "Toward a Critical Phenomenology of Illegality: State Power, Criminalization, and Abjectivity among Undocumented Migrant Workers in Tel Aviv, Israel." *International Migration* 45 (3): 8–38.

Yang, Anand A. 1998. *Bazaar India: Markets, Society, and the Colonial State in Gangetic Bihar.* Berkeley: University of California Press.

Zee Media Bureau. 2013. "Lord Kedarnath Moved to Ukhimath: Right or Wrong." *Znews,* June 26. http://zeenews.india.com/news/uttarakhand/lord-kedarnath-moved-to-ukhimath-%E2%80%93-right-or-wrong_857868.html.

Zee Media Bureau and Press Trust of India. 2013. "Uttarakhand Floods: Death Toll Uncertain, Minister Says 10,000 Just an Estimate." *Znews,* June 30. http://zeenews.india.com/news/nation/uttarakhand-floods-death-toll-uncertain-minister-says-10000-just-an-estimate_858805.html.

Zoloth, Laurie. 2015. "Risky Hospitality: Ordinal Ethics and the Duties of Abundance." *Journal of the American Academy of Religion* 83 (2): 373–87. https://doi.org/10.1093/jaarel/lfv014.

INDEX

Page numbers in italic refer to figures, maps, and tables. Thus, *122f10* refers to figure 10 on page *122*; *xv*map *1* refers to map 1 on page *xv*; *102t2* refers to table 2 on page *102*. Numbers followed by an "n" or "nn" refer to notes.

Acharya, Diwakar, 61, 62, 63
acheri. See supernatural beings
Agnihotri, Rajesh, 226n17
Agrawal, Arun, and K. Sivaramakrishnan, 86, 179, 186
Amarnath: attraction (Hindi: *akarshan*) of, 17; as a *dham* of Shiva, 7
apda (disaster): and the "broader situation," 145–146, 155–156, 168; and *dosh*, 187–188, 196; and the eco-social system, 187–188, 193–196; flooding in 2013 in Uttarakhand, 8–10, *9t1*, 27–28, 145–167; flooding in Rambara described in *Apda Ka Kafan*, 149; glacial lake outburst floods (GLOFs), 99, 169, 193; mountain hazards, 3, 225n3; and the resumption of the Kedarnath *doli*, 164, 173; and scholars of disaster, 146
Atimarga ("Outer Path"), 62, 63–64

Badrinath, 123, 136; closing for the winter of, 164; during the flood, 150; glossed, 211; location in Uttarakhand, *xv*map *1*; Nivedita's visit to, 96; as one of the four abodes of the Uttarakhand Char Dham Yatra, 2, 86, 199, 200; as part

of British Garhwal, 88, 94–95; and royal patronage, 69; and the Shri Badrinath and Kedarnath Temples Act of 1939, 93–94; sightings of deities at, 131; in the twelfth to sixteenth centuries, 87; *yatra* tourism to, 101–103, *102t2*; and *yatra* trade, 87. *See also* Samiti (Badrinath-Kedarnath Temples Committee)
Badrinath-Kedarnath Temples Committee. *See* Samiti
Bahuguna, Sunderlal: and the Chipko movement, 90; "ecology is permanent economy" pronounced by, 195; remarks on sustainability, 207–208
Bakker, Hans, 58
Banerjee, Sandeep, and Subho Basu, 185–186
Bauer, Andrew M., and Mona Bhan, 24
Berry, Evan, 202
Berti, Daniela, 204–205, 222n4, 223n4
Bhairav(a), Bhairavnath, 53, 108; anti-*vikas* position of, 191; devotion to, 20; and evening *arati* at Kedarnath, 67, 125; as the lineage deity of the Kedarnath Tirth Purohit Association, 65; local forms of, 53; as an official province in the off season, 112; relationship to Shiva, 64–65; and Shaiva Tantra, 67; sightings of Kal Bhairav, 131. *See also* Bhukund Bhairavnath
Bhardwaj, Surinder Mohan, 46, 113
Bhardwaj, Surinder Mohan, and J. G. Lochtefeld, 19

Milton Keynes UK
Ingram Content Group UK Ltd.
UKHW022003061123
432089UK00011B/105